Applied Mathematical Sciences

EDITORIAL STATEMENT

The mathematization of all sciences, the fading of traditional scientific boundaries, the impact of computer technology, the growing importance of mathematical-computer modelling and the necessity of scientific planning all create the need both in education and research for books that are introductory to and abreast of these developments.

The purpose of this series is to provide such books, suitable for the user of mathematics, the mathematician interested in applications, and the student scientist. In particular, this series will provide an outlet for material less formally presented and more anticipatory of needs than finished texts or monographs, yet of immediate interest because of the novelty of its treatment of an application or of mathematics being applied or lying close to applications.

The aim of the series is, through rapid publication in an attractive but inexpensive format, to make material of current interest widely accessible. This implies the absence of excessive generality and abstraction, and unrealistic idealization, but with quality of exposition as a goal.

Many of the books will originate out of and will stimulate the development of new undergraduate and graduate courses in the applications of mathematics. Some of the books will present introductions to new areas of research, new applications and act as signposts for new directions in the mathematical sciences. This series will often serve as an intermediate stage of the publication of material which, through exposure here, will be further developed and refined. These will appear in conventional format and in hard cover.

MANUSCRIPTS

The Editors welcome all inquiries regarding the submission of manuscripts for the series. Final preparation of all manuscripts will take place in the editorial offices of the series in the Division of Applied Mathematics, Brown University, Providence, Rhode Island.

SPRINGER-VERLAG NEW YORK INC., 175 Fifth Avenue, New York, N. Y. 10010

Printed in U.S.A.

Applied Mathematical Sciences | Volume 46

Applied Mathematical Sciences

1. John: **Partial Differential Equations**, 4th ed. (cloth)
2. Sirovich: **Techniques of Asymptotic Analysis.**
3. Hale: **Theory of Functional Differential Equations**, 2nd ed. (cloth)
4. Percus: **Combinatorial Methods.**
5. von Mises/Friedrichs: **Fluid Dynamics.**
6. Freiberger/Grenander: **A Short Course in Computational Probability and Statistics.**
7. Pipkin: **Lectures on Viscoelasticity Theory.**
8. Giacaglia: **Perturbation Methods in Non-Linear Systems.**
9. Friedrichs: **Spectral Theory of Operators in Hilbert Space.**
10. Stroud: **Numerical Quadrature and Solution of Ordinary Differential Equations.**
11. Wolovich: **Linear Multivariable Systems.**
12. Berkovitz: **Optimal Control Theory.**
13. Bluman/Cole: **Similarity Methods for Differential Equations.**
14. Yoshizawa: **Stability Theory and the Existence of Periodic Solutions and Almost Periodic Solutions.**
15. Braun: **Differential Equations and Their Applications**, 3rd ed. (cloth)
16. Lefschetz: **Applications of Algebraic Topology.**
17. Collatz/Wetterling: **Optimization Problems.**
18. Grenander: **Pattern Synthesis: Lectures in Pattern Theory, Vol I.**
19. Marsden/McCracken: **The Hopf Bifurcation and its Applications.**
20. Driver: **Ordinary and Delay Differential Equations.**
21. Courant/Friedrichs: **Supersonic Flow and Shock Waves.** (cloth)
22. Rouche/Habets/Laloy: **Stability Theory by Liapunov's Direct Method.**
23. Lamperti: **Stochastic Processes: A Survey of the Mathematical Theory.**
24. Grenander: **Pattern Analysis: Lectures in Pattern Theory, Vol. II.**
25. Davies: **Integral Transforms and Their Applications.**
26. Kushner/Clark: **Stochastic Approximation Methods for Constrained and Unconstrained Systems.**
27. de Boor: **A Practical Guide to Splines.**
28. Keilson: **Markov Chain Models—Rarity and Exponentiality.**
29. de Veubeke: **A Course in Elasticity.**
30. Sniatycki: **Geometric Quantization and Quantum Mechanics.**
31. Reid: **Sturmian Theory for Ordinary Differential Equations.**
32. Meis/Markowitz: **Numerical Solution of Partial Differential Equations.**
33. Grenander: **Regular Structures: Lectures in Pattern Theory, Vol. III.**
34. Kevorkian/Cole: **Perturbation Methods in Applied Mathematics.** (cloth)
35. Carr: **Applications of Centre Manifold Theory.**

(continued after Index)

Calvin H. Wilcox

Scattering Theory for Diffraction Gratings

Springer-Verlag
New York Berlin Heidelberg Tokyo

Calvin H. Wilcox
University of Utah
Department of Mathematics
Salt Lake City, Utah 84112
U.S.A.

AMS Classification: 34B99, 78A45, 76B15

Library of Congress Cataloging in Publication Data
Wilcox, Calvin H. (Calvin Hayden)
Scattering theory for diffraction gratings.
(Applied mathematical sciences ; v. 46)
Bibliography: p.
Includes index.
1. Diffraction gratings. 2. Scattering (Mathematics)
I. Title. II. Series: Applied mathematical sciences
(Springer-Verlag New York Inc.) ; v.. 46.
QA1.A647 vol. 46 [QC417] 510s [530.1'5] 83-19976

With 4 illustrations

Printed and bound by R. R. Donnelley & Sons, Harrisonburg, Virginia.
Printed in the United States of America.

9 8 7 6 5 4 3 2 1

ISBN 0-387-90924-9 Springer-Verlag New York Berlin Heidelberg Tokyo
ISBN 3-540-90924-9 Springer-Verlag Berlin Heidelberg New York Tokyo

Preface

The scattering of acoustic and electromagnetic waves by periodic surfaces plays a role in many areas of applied physics and engineering. Optical diffraction gratings date from the nineteenth century and are still widely used by spectroscopists. More recently, diffraction gratings have been used as coupling devices for optical waveguides. Trains of surface waves on the oceans are natural diffraction gratings which influence the scattering of electromagnetic waves and underwater sound. Similarly, the surface of a crystal acts as a diffraction grating for the scattering of atomic beams. This list of natural and artificial diffraction gratings could easily be extended.

The purpose of this monograph is to develop from first principles a theory of the scattering of acoustic and electromagnetic waves by periodic surfaces. In physical terms, the scattering of both time-harmonic and transient fields is analyzed. The corresponding mathematical model leads to the study of boundary value problems for the Helmholtz and d'Alembert wave equations in plane domains bounded by periodic curves. In the formalism adopted here these problems are intimately related to the spectral analysis of the Laplace operator, acting in a Hilbert space of functions defined in the domain adjacent to the grating.

The intended audience for this monograph includes both those applied physicists and engineers who are concerned with diffraction gratings and those mathematicians who are interested in spectral analysis and scattering theory for partial differential operators. An attempt to address simultaneously two such disparate groups must raise the question: is there a common domain of discourse? The honest answer to this question is no! Current mathematical literature on spectral analysis and scattering theory is based squarely on functional analysis, particularly the theory of linear transformations in Hilbert spaces. This theory has been readily accessible ever since the publication of M. H. Stone's AMS Colloquium volume in 1932. Nevertheless, the theory has not become a part of the curricula of applied physics and engineering and it is seldom seen in applied science literature on wave propagation and scattering. Instead, that literature is characterized by, on the one hand, the use of heuristic non-rigorous arguments and, on the other, by formal manipulations that typically involve divergent series and integrals, generalized functions of unspecified types and the like.

The differences in style and method outlined above pose a dilemma. Can an exposition of our subject be written that is accessible and useful to both applied scientists and mathematicians? An attempt is made to do this below by dividing the work into two parts. Part 1, called Physical Theory, presents the basic physical concepts and results, formulated in the simplest and most concise form consistent with their nature. Moreover, Part 1 can be interpreted in two different ways. First, it can be interpreted in the heuristic way favored by applied physicists and engineers. When read in this way it presents a complete statement of the physical content of the theory. Second, for readers conversant with Hilbert space theory Part 1 can be interpreted as a concise statement of the principal concepts and results of a rigorous mathematical theory.

When read in the second way Part 1 serves as an introduction to and overview of the complete theory which is presented in Part 2, Mathematical Theory. This part develops the relevant concepts and results from functional analysis and the theory of partial differential equations and applies them to give complete proofs of the results formulated in Part 1. At the same time many secondary concepts and results are formulated and proved that lead to a deeper understanding of the nature and limitations of the theory.

Acknowledgments

Preliminary studies for this work began in 1974, while I was a visiting professor at the University of Stuttgart, and continued there during my tenure as an Alexander von Humboldt Foundation Senior Scientist in 1976-77. The work was completed during my sabbatical year in 1980 when I was a visiting professor at the University of Bonn with the support of the Sonderforschungsbereich 72. Throughout this period my research was supported by the U. S. Office of Naval Research. I should like to express here my appreciation for the support of the Universities of Bonn, Stuttgart and Utah, the Alexander von Humboldt Foundation and the Office of Naval Research which made the work possible. My special thanks are expressed to Professor Rolf Leis, University of Bonn, and Professor Peter Werner, University of Stuttgart, for arranging my visits to their universities.

I should also like to thank here Professor Jean Claude Guillot of the University of Paris for helpful discussions in 1977-78 of a preliminary version of this work. Finally, and most important of all, I want to express my thanks to Dr. Hans Dieter Alber of the University of Bonn for his outstanding paper of 1979 on steady-state scattering by periodic surfaces. It was the concepts introduced in this paper that led to the final, very general, theory developed here. Dr. Alber's contributions have influenced nearly every part of this work.

Calvin H. Wilcox
Bonn
July, 1982

Contents

Introduction

The first theoretical studies of scattering by diffraction gratings are due to Lord Rayleigh. His "Theory of Sound" Volume 2, 2nd Edition, published in 1896 [18]*, contains an analysis of the scattering of a monochromatic plane wave normally incident on a grating with a sinusoidal profile. In a subsequent paper [19] he extended the analysis to oblique incidence. Rayleigh assumed in his work that in the half-space above the grating the reflected wave is a superposition of the specularly reflected plane wave, a finite number of secondary plane waves propagating in the directions of the higher order grating spectra of optics, and an infinite sequence of evanescent waves whose amplitudes decrease exponentially with distance from the grating. The validity of Rayleigh's assumption for general grating profiles was realized in the early 1930's [10], following Bloch's work [4] on the analogous problem of de Broglie waves in crystals. Waves of this type will be called Rayleigh-Bloch waves (R-B waves for brevity) in this work.

The goal of Rayleigh's work and the literature based on it was to calculate the relative amplitudes and phases of the diffracted plane wave components of the R-B waves. Several methods for doing this have been developed. L. A. Weinstein [27] and J. A. DeSanto [5,6] gave exact solutions to the problem of the scattering of monochromatic plane waves by a comb grating; i.e., an array of periodically spaced infinitesimally thin parallel plates of finite depth mounted perpendicularly on a plane. For gratings with sinusoidal profiles, infinite systems of linear equations for the complex reflection coefficients were given by J. L. Uretsky [26] and J. A. DeSanto [7]. More recently, DeSanto [8] has extended his results to essentially arbitrary profiles. Finally, an excellent review up to 1980 of

*Numbers in square brackets denote references from the list at the end of the monograph.

both theoretical and numerical methods for determining the reflection coefficients is contained in the book Electromagnetic Theory of Gratings, edited by R. Petit [17].

The literature on diffraction gratings and their applications is very large. References to work done before 1967 may be found in the monograph by Stroke in the Handbuch der Physik [25]. A survey of the literature up to 1980 is contained in [17].

The works referenced above provide a satisfactory understanding of the scattering of the steady beams used in classical spectroscopy. However, modern applications of gratings in such areas as optical waveguides and underwater sound require an understanding of how transient electromagnetic and acoustic fields, such as pulsed laser beams and sonar signals, are scattered by diffraction gratings. The existing grating theories are inadequate for the analysis of these problems.

The purpose of this monograph is to develop a theory of the scattering of transient electromagnetic and acoustic fields by diffraction gratings. The theory is based on an eigenfunction expansion for gratings in which the eigenfunctions are R-B waves. The analysis parallels the author's work on the scattering of transient sound waves by bounded obstacles [30,31,33]. The eigenfunction expansions are generalizations of T. Ikebe's theory of distorted plane wave expansions [12], first developed for quantum mechanical potential scattering and subsequently extended to a variety of scattering problems [2,15,21,22,23,32]. The theory is based on the study of a linear operator A, called here the grating propagator, which is a selfadjoint realization of the negative of the Laplace operator acting in the Hilbert space of square integrable acoustic fields. A fundamental result of this analysis is a representation of the spectral family of A by means of R-B waves. The R-B wave expansions follow as a corollary.

The theory of scattering by gratings developed below is restricted, for brevity, to the case of two-dimensional wave propagation. Specifically, the waves are assumed to be solutions of the wave equation in a two-dimensional grating domain and to satisfy the Dirichlet or Neumann boundary condition on the grating profile. These problems provide models for the scattering of sound waves by acoustically soft or rigid gratings and of TE or TM electromagnetic waves by perfectly conducting gratings. It will be seen that the methods employed are also applicable to the scattering of scalar waves by three-dimensional (and n-dimensional) gratings and to systems such as Maxwell's equations and the equations of elasticity.

Even with the restriction to the two-dimensional case, the analytical work needed to derive and fully establish eigenfunction expansions for diffraction gratings is necessarily intricate and lengthy. This is clear from an examination of the simpler case of scattering by bounded obstacles presented in the author's monograph [30]. Therefore, to make the work more accessible to potential users, the monograph has been divided into two parts. As explained in the Preface, Part 1 can be interpreted both as a complete statement, without proofs, of the physical concepts and results of the theory and also as a summary and introduction to the complete mathematical theory developed in Part 2.

A preliminary version of the R-B wave expansion theorem of this monograph was announced by J. C. Guillot and the author in 1978 [34]. That work was based on an integral equation for the R-B waves. In this monograph an alternative method based on analytic continuation is used. A key step is the introduction of the reduced grating propagator A_p which depends on the wave momentum. The Hilbert space theory of such operators was initiated by H. D. Alber [3]. Alber's powerful method of analytic continuation of the resolvent of A_p is used in Part 2 to construct the R-B wave eigenfunctions.

Part 1
Physical Theory

This monograph develops a theory of the scattering of two-dimensional acoustic and electromagnetic fields by diffraction gratings. This Part 1 presents the principal physical concepts and results in their simplest forms and without proofs. Moreover, to avoid distracting technicalities the precise conditions for the validity of the results are not always given. Part 1 also contains no references to the literature. All of these omissions are remedied in Part 2 which contains the final mathematical formulation of the theory, together with complete proofs and indications of related literature.

§1. The Physical Problems

The propagation of two-dimensional acoustic and electromagnetic fields is studied below in unbounded planar regions whose boundaries (= the diffraction gratings) lie between two parallel lines and are periodic. In each case the medium filling the region is assumed to be homogeneous and lossless. In the acoustic case the grating is assumed to be either rigid or acoustically soft. In the electromagnetic case it is assumed to be perfectly conducting. In both cases the sources of the field are assumed to be localized in space and time. The principal goal of the theory is to calculate the "final" or large-time form of the resulting transient field.

§2. The Mathematical Formulation

Rectangular coordinates $X = (x,y) \in R^2$ will be used to describe the region adjacent to the diffraction grating. The notation

$$R_a^2 = \{X : y > a\} \tag{2.1}$$

will be used. Then with a suitable choice of coordinate axes the region

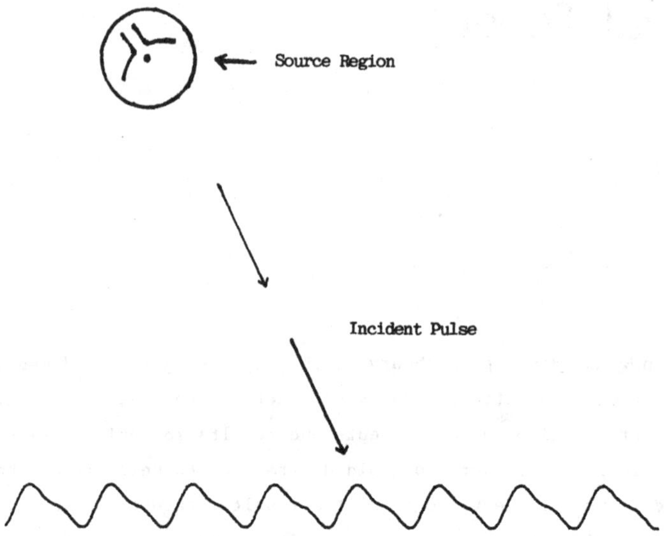

Figure 1. Grating with Source Region and Incident Pulse

above the grating will be characterized by a grating domain $G \subset R^2$ with the properties

$$R_h^2 \subset G \subset R_0^2 , \tag{2.2}$$

$$G + (2\pi,0) = G \tag{2.3}$$

where $h > 0$ is a suitable constant. The choice of the constant 2π in (2.3) is simply a convenient normalization.

The acoustic or electromagnetic field in G can be described by a real-valued function $u = u(t,X)$ that is a solution of the initial-boundary value problem

$$D_t^2 u - \Delta u = 0 \text{ for all } t > 0 \text{ and } X \in G , \tag{2.4}$$

$$D_\nu u \equiv \vec{\nu} \cdot \nabla u = 0 \text{ (resp., } u = 0) \text{ for all } t \geq 0 \text{ and } X \in \partial G , \tag{2.5}$$

$$u(0,X) = f(X) \text{ and } D_t u(0,X) = g(X) \text{ for all } X \in G . \tag{2.6}$$

Here t is a time coordinate, $D_t = \partial/\partial t$, $D_x = \partial/\partial x$, $D_y = \partial/\partial y$, $\nabla u = (D_x u, D_y u)$, $\Delta u = D_x^2 u + D_y^2 u$, ∂G denotes the boundary of G and $\vec{\nu} = \vec{\nu}(X)$ is a unit normal vector to ∂G at X. In the acoustic case $u(t,X)$ is interpreted as a potential for an acoustic field with velocity $\vec{v} = -\nabla u$ and acoustic pressure

$p = D_t u$. Then the boundary condition (2.5) corresponds to an acoustically hard (resp., soft) boundary. Alternatively, if u satisfies the Neumann condition $D_\nu u = 0$ on ∂G then

$$E_x = D_y u,\ E_y = -D_x u,\ H_z = D_t u \tag{2.7}$$

describes a TM electromagnetic field in a domain G bounded by a perfect electrical conductor. Similarly, if u satisfies the Dirichlet condition $u = 0$ on ∂G then

$$H_x = -D_y u,\ H_y = D_x u,\ E_z = D_t u \tag{2.8}$$

describes a TE electromagnetic field in the same kind of domain. The functions f(X) and g(X) in (2.6) characterize the initial state of the field. They are assumed to be given or calculated from the prescribed wave sources, and to be localized:

$$\operatorname{supp} f \cup \operatorname{supp} g \subset \{X : x^2 + (y - y_0)^2 \le \delta_0^2\} \tag{2.9}$$

where $y_0 > h + \delta_0$.

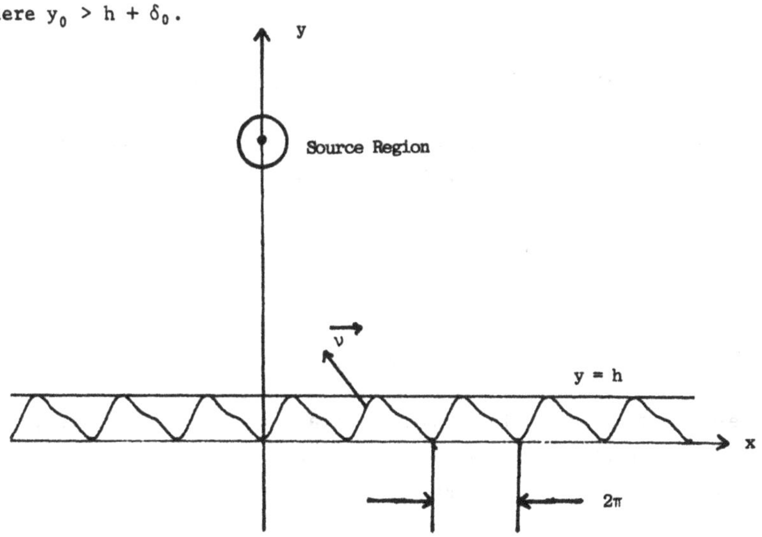

Figure 2. Grating Domain with Coordinate System

In both the acoustic and the electromagnetic interpretations the integral

$$E(u,K,t) = \int_K \{|\nabla u(t,X)|^2 + |D_t u(t,X)|^2\}\, dX \tag{2.10}$$

is interpreted as the wave energy in the set K at time t (dX = dxdy). Solutions of the wave equation satisfy the energy conservation law E(u,G,t) = E(u,G,0) under both boundary conditions (2.5). It will be assumed that the initial state has finite energy:

$$\int_G \{|\nabla f(X)|^2 + |g(X)|^2\}\, dX < \infty \,. \tag{2.11}$$

§3. Solution of the Initial-Boundary Value Problem

The initial-boundary value problem in its classical formulation (2.4)-(2.6) will have a solution only if ∂G and the functions f and g are sufficiently smooth. However, for arbitrary domains G the problem is known to have a unique solution with finite energy whenever the initial state f,g has this property. A formal construction of the solution may be based on the linear operator $A = -\Delta$, acting in the Hilbert space $\mathcal{H} = L_2(G)$. If the domain of A is defined to be the set of $u \in \mathcal{H}$ such that $\nabla u \in \mathcal{H}$, $\Delta u \in \mathcal{H}$ and one of the boundary conditions (2.5) holds then A is a selfadjoint non-negative operator. Moreover,

$$u(t,\cdot) = (\cos t\, A^{1/2})\, f + (A^{-1/2} \sin t\, A^{1/2})\, g \tag{3.1}$$

is the solution with finite energy whenever the initial state has finite energy. It will be convenient to write (3.1) as

$$u(t,X) = \mathrm{Re}\,\{v(t,X)\}\,, \quad v(t,\cdot) = e^{-itA^{1/2}}\, h \tag{3.2}$$

where

$$h = f + i\, A^{-1/2}\, g\,. \tag{3.3}$$

This representation is valid if f and g are real-valued and $A^{1/2} f$, f, g and $A^{-1/2} g$ are in $\mathcal{H}$. A rigorous interpretation of relations (3.1)-(3.3) can be based on the calculus of selfadjoint operators in Hilbert spaces.

§4. The Reference Problem and Its Eigenfunctions

In the class of grating domains defined by (2.2), (2.3) there is a special case for which the scattering problem is explicitly solvable. This is the case of the degenerate grating $G = R_0^2$ ($h = 0$). The problem (2.4)-(2.6) with $G = R_0^2$ and the Neumann boundary condition will be called the reference problem. The corresponding reference propagator is the operator $A_0 = -\Delta$ in $\mathcal{H}_0 = L_2(R_0^2)$ with Neumann boundary condition. The solution of the scattering problem for non-degenerate gratings is developed below as a perturbation of the reference problem.

A_0 has a pure continuous spectrum filling the half-line $\lambda \geq 0$. This is easily verified by separation of variables which yields the family of generalized eigenfunctions

$$(4.1) \qquad \psi_0(x,y,p,q) = \frac{1}{\pi} e^{ipx} \cos q\, y = \frac{1}{2\pi} e^{i(px-qy)} + \frac{1}{2\pi} e^{i(px+qy)}$$

where $(x,y) \in R_0^2$ and also $(p,q) \in R_0^2$. Clearly

$$(4.2) \qquad A_0\, \psi_0(x,y,p,q) = -\Delta\, \psi_0(x,y,p,q) = \omega^2(p,q)\, \psi_0(x,y,p,q)$$

where

$$(4.3) \qquad \omega^2(p,q) = p^2 + q^2 \geq 0$$

and

$$(4.4) \qquad D_\nu\, \psi_0(x,y,p,q) = D_y\, \psi_0(x,y,p,q) = 0 \text{ on } \partial R_0^2 \,.$$

The decomposition of (4.1) illustrates the physical interpretation of ψ_0. If $(p,q) \in R_0^2$ then $q > 0$ and the first term represents a monochromatic plane wave incident on the plane boundary in the direction $(p,-q)$, while the second term represents the specularly reflected wave propagating in the direction (p,q).

The functions $\{\psi_0(X,P) : P = (p,q) \in R_0^2\}$ form a complete family of generalized eigenfunctions for A_0. This means that for every $h \in \mathcal{H}_0$ one has

$$(4.5) \qquad \hat{h}_0(P) = \ell.i.m. \int_{R_0^2} \overline{\psi_0(X,P)}\, h(X)\, dX \text{ exist in } \mathcal{H}_0$$

and

$$(4.6) \qquad h(X) = \ell.i.m. \int_{R_0^2} \psi_0(X,P)\, \hat{h}_0(P)\, dP \text{ in } \mathcal{H}_0 \,.$$

The $\ell.i.m.$ notation in (4.5) means that the integral converges not pointwise but in $\mathcal{H}_0$; i.e.,

$$(4.7) \qquad \lim_{M\to\infty} \int_{R_0^2} \left| \hat{h}_0(P) - \int_0^M \int_{-M}^M \overline{\psi_0(X,P)}\, h(X)\, dX \right|^2 dP = 0 \,.$$

Equation (4.6) has the analogous interpretation. Moreover, Parseval's relation holds:

(4.8) $$\int_{R_0^2} |\hat{h}_0(P)|^2 \, dP = \int_{R_0^2} |h(X)|^2 \, dX \, .$$

In fact, if a linear operator Φ_0 in $\mathcal{H}_0$ is defined by

(4.9) $$\Phi_0 h = \hat{h}_0$$

then Φ_0 is unitary.

The eigenfunction expansion (4.6) is useful because it diagonalizes A_0. In particular, the solution $v_0(t,\cdot) = e^{-itA_0^{1/2}} h$ of the reference problem has the expansion

(4.10) $$v_0(t,X) = \ell.\text{i.m.} \int_{R_0^2} \psi_0(X,P) \, e^{-it\omega(P)} \, \hat{h}_0(P) \, dP$$

where $\omega(P) = |P| = \sqrt{p^2+q^2}$.

§5. Rayleigh-Bloch Diffracted Plane Waves for Gratings

In analogy with the case of the degenerate grating, the generalized eigenfunctions of the grating propagator A may be defined as the response of the grating to a monochromatic plane wave $(2\pi)^{-1} \exp\{i(px - qy)\}$. It will be shown that there are two distinct families which will be denoted by $\psi_+(X,P)$ and $\psi_-(X,P)$, respectively. It will be convenient to write them as perturbations of the eigenfunctions $\psi_0(X,P)$ for the degenerate grating:

(5.1) $$\psi_\pm(X,P) = \psi_0(X,P) + \psi_\pm^{sc}(X,P) \, , \quad X \in G \, , \quad P \in R_0^2 \, .$$

They are characterized by the conditions

(5.2) $$A \, \psi_\pm(X,P) = -\Delta \, \psi_\pm(X,P) = \omega^2(P) \, \psi_\pm(X,P) \, , \quad X \in G \, ,$$

(5.3) $$D_\nu \, \psi_\pm = 0 \ (\text{resp.}, \ \psi_\pm = 0) \ \text{for } X \in \partial G \, ,$$

(5.4) $\psi_+^{sc}(X,P)$ is outgoing and $\psi_-^{sc}(X,P)$ is incoming for $X \to \infty$.

The last condition is based on the Fourier series representation in x of $\psi_\pm^{sc}(x,y,P)$ which is valid for $y > h$. It can be written (with $P = (p,q)$)

(5.5) $$\begin{aligned} \psi_\pm^{sc}(X,P) = {} & \frac{1}{2\pi} \sum_{(p+\ell)^2<p^2+q^2} c_\ell^\pm(P) \, e^{i(p_\ell x \pm q_\ell y)} \\ & + \frac{1}{2\pi} \sum_{(p+\ell)^2 \geq p^2+q^2} c_\ell^\pm(P) \, e^{ip_\ell x} \, e^{-y\{(p+\ell)^2-p^2-q^2\}^{1/2}} \end{aligned}$$

where

$$(5.6) \qquad p_\ell = p + \ell \, , \quad q_\ell = (p^2 + q^2 - (p + \ell)^2)^{1/2}$$

and the summations in (5.5) are over the integers ℓ that satisfy the indicated inequalities. Note that the first sum is finite. Moreover, if $P_\ell = (p_\ell, q_\ell)$ then

$$(5.7) \qquad \omega^2(P_\ell) = p_\ell^2 + q_\ell^2 = p^2 + q^2 = \omega^2(P)$$

which asserts that the wave frequency is preserved under scattering. It is clear from (5.5) that the families ψ_+ and ψ_- are distinct. In fact, one family can be obtained from the other by means of the identity

$$(5.8) \qquad \psi_-(X,p,q) = \overline{\psi_+(X,-p,q)} \; .$$

(5.1) and (5.5) imply that ψ_+ describes the outgoing response to the incoming plane wave $(2\pi)^{-1} \exp\{i(px - qy)\}$ while ψ_- describes the incoming response to the outgoing plane wave $(2\pi)^{-1} \exp\{i(px + qy)\}$. ψ_+ and ψ_- will be called, respectively, the outgoing and incoming Rayleigh-Bloch diffracted plane waves for G.

§6. Rayleigh-Bloch Surface Waves for Gratings

The terms in the second sum of (5.5) may be called surface waves for the grating since they propagate in the x-direction, parallel to the grating, and are damped exponentially with distance y from the grating. These waves are driven by the incident wave $(2\pi)^{-1} \exp\{i(px \mp qy)\}$. It can happen that there exist certain curves

$$(6.1) \qquad \lambda_j(p) = \omega^2(p,q)$$

and corresponding functions $\psi_j(X,p)$ such that

$$(6.2) \qquad A\, \psi_j(X,p) = -\Delta\, \psi_j(X,p) = \lambda_j(p)\, \psi_j(X,p) \; , \quad X \in G \; ,$$

$$(6.3) \qquad D_\nu\, \psi_j(X,p) = 0 \text{ (resp., } \psi_j(X,p) = 0) \text{ for } X \in \partial G \; ,$$

and ψ_j is a pure surface wave; that is, for $y > h$ one has

$$(6.4) \qquad \psi_j(X,p) = \sum_{p_\ell^2 > \lambda_j(p)} c_{j\ell}(p)\, e^{ip_\ell x}\, e^{-y\{p_\ell^2 - \lambda_j(p)\}^2} \; .$$

In the Dirichlet case it is known that if ∂G is a single smooth curve

$y = h(x)$ then no such R-B surface waves exist. No general criteria are known in the Neumann case.

In the remainder of Part 1 it is assumed for simplicity that G admits no R-B surface waves. The modifications needed when there are R-B surface waves are discussed in Part 2.

§7. Rayleigh-Bloch Wave Expansions

The R-B diffracted plane wave expansions for G are exactly analogous to that for the reference problem of §4. Thus each of the families $\{\psi_+(X,P) : P \in R_0^2\}$ and $\{\psi_-(X,P) : P \in R_0^2\}$ is a complete family of generalized eigenfunctions for A. This means that for every $h \in \mathcal{H} = L_2(G)$ one has

$$(7.1) \qquad \hat{h}_\pm(P) = \ell.\text{i.m.} \int_G \overline{\psi_\pm(X,P)}\, h(X)\, dX$$

exists in $\mathcal{H}_0 = L_2(R_0^2)$ and

$$(7.2) \qquad h(X) = \ell.\text{i.m.} \int_{R_0^2} \psi_\pm(X,P)\, \hat{h}_\pm(P)\, dP$$

in $\mathcal{H}$. Moreover, Parseval's relation holds:

$$(7.3) \qquad \int_{R_0^2} |\hat{h}_\pm(P)|^2\, dP = \int_G |h(X)|^2\, dX$$

and the linear operators Φ_+ and Φ_- from $\mathcal{H}$ to $\mathcal{H}_0$ defined by

$$(7.4) \qquad \Phi_\pm h = \hat{h}_\pm$$

are unitary. Finally, the representation diagonalizes A, just as in the case of the reference problem. In particular the solution $v(t,\cdot) = e^{-itA^{1/2}} h$ of the scattering problem has the two expansions

$$(7.5) \qquad v(t,X) = \ell.\text{i.m.} \int_{R_0^2} \psi_\pm(X,P)\, e^{-it\omega(P)}\, \hat{h}_\pm(P)\, dP .$$

The ψ_+ and ψ_- expansions will be called the outgoing and incoming expansions, respectively. It will be shown that both are useful in determining the structure of the scattered field.

§8. Wave and Scattering Operators for Gratings

The basic program of time-dependent scattering theory is to show that each evolution $v(t) = \exp\{-itA^{1/2}\}h$ of a given system is asymptotically equal, for $t \to \infty$, to an evolution $v_0(t) = \exp\{-itA_0^{1/2}\}h_{sc}$ of a simpler "reference system." This means that

(8.1) $$\lim_{t\to\infty} \left\| J\, e^{-itA^{1/2}} h - e^{-itA_0^{1/2}} h_{sc} \right\| = 0$$

where $J : \mathcal{H} \to \mathcal{H}_0$ is a suitable bounded linear operator.

In Part 2 the R-B wave expansions are used to demonstrate the behavior (8.1) for grating domains G that admit no surface waves. The reference domain is the degenerate grating R_0^2 and

(8.2) $$J\, h(X) = \begin{cases} h(X), & X \in G , \\ 0 , & X \in R_0^2 - G . \end{cases}$$

It is easy to see that the asymptotic state h_{sc}, when it exists, is uniquely determined by (8.1). In fact, h_{sc} is related to h by $(h_{sc})\hat{}_0 = \hat{h}_-$; i.e., $\Phi_0 h_{sc} = \Phi_- h$ or

(8.3) $$h_{sc} = \Phi_0^* \Phi_- h .$$

This relationship is well known in applications of scattering theory to both quantum and classical physics.

Condition (8.1) is equivalent to the existence of the wave operator W_+ where

(8.4) $$W_\pm = \text{s-}\lim_{t\to\pm\infty} e^{itA_0^{1/2}} J\, e^{-itA^{1/2}}$$

and s-$\lim$ denotes strong convergence. Moreover, (8.3) and the analogues of (8.1), (8.3) for $t \to -\infty$ imply that

(8.5) $$W_\pm = \Phi_0^* \Phi_\mp .$$

It follows from (8.5) and the results of §7 that W_+ and W_- are unitary operators from $\mathcal{H}$ to $\mathcal{H}_0$.

The scattering operator S of the abstract theory of scattering is defined by

(8.6) $$S = W_+ W_-^* .$$

The unitarity of the wave operators implies that S is a unitary operator in $\mathcal{H}_0$. The related operator

(8.7) $$\hat{S} = \Phi_- \Phi_+^* ,$$

often called the Heisenberg operator, or S-matrix, is also unitary in $\mathcal{H}_0$. (8.5), (8.6) and (8.7) imply that

$$S = \Phi_0^* \hat{S} \Phi_0 . \tag{8.8}$$

Relations (8.4)-(8.8) are standard definitions and results in the abstract theory of scattering. An explicit construction of the S-matrix of a diffraction grating, based on the R-B waves, is described next. It is applied in §10 below to describe the scattering by gratings of signals from remote sources.

The S-matrix is determined by the scattering coefficients $c_\ell^-(P)$ of the R-B wave $\psi_-(X,P)$ defined by (5.5). To describe the structure of $\hat{S}$ it will be convenient to define the following sets in the space of the momentum variables $P = (p,q)$.

$$\pi_\ell = R_0^2 \cap \{(p,q) : q_\ell^2 \equiv p^2 + q^2 - (p + \ell)^2 = 0\} . \tag{8.9}$$

Clearly, for $\ell \neq 0$, π_ℓ is the portion in R_0^2 of the parabola with focus (0,0) and vertex $(-\ell/2,0)$. Next, define

$$\tag{8.10} \begin{cases} \mathcal{O}_m = \text{domain between } \pi_m \text{ and } \pi_{m+1}, & m = 0,1,2,\cdots \\ \mathcal{O}_{-n} = \text{domain between } \pi_{-n} \text{ and } \pi_{-n-1}, & n = 0,1,2,\cdots \\ \mathcal{O}_{m,n} = \mathcal{O}_m \cap \mathcal{O}_{-n}, \quad m,n = 0,1,2,\cdots \end{cases}$$

Examination of the Fourier expansion (5.5) of $\psi_\pm$ shows that $P \in \mathcal{O}_{m,n}$ if and only if $\psi_\pm(X,P)$ contains exactly $m+n+1$ outgoing (for ψ_+) or incoming (for ψ_-) plane waves with the propagation vectors (p_ℓ, q_ℓ), $-n \le \ell \le m$.

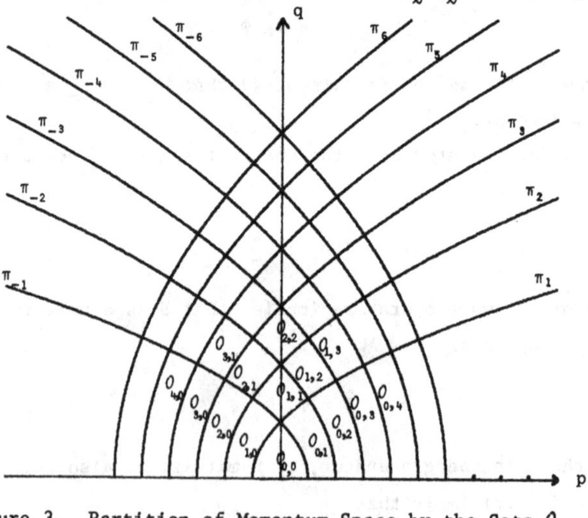

Figure 3. Partition of Momentum Space by the Sets $\mathcal{O}_{m,n}$

Next let $\ell \in Z$ (the integers) and consider the mappings X_ℓ of the momentum space R_0^2 defined by

$$D(X_\ell) = \{(p,q) : \sqrt{p^2+q^2} > |p+\ell|,\ q > 0\} \tag{8.11}$$

and

$$X_\ell(p,q) = (p_\ell, q_\ell) = (p+\ell, \{p^2+q^2-(p+\ell)^2\}^{1/2}) \tag{8.12}$$

X_0 is the identity map. For $\ell \in Z - \{0\}$, X_ℓ is analytic on its domain and maps it bijectively onto the range

$$R(X_\ell) = D(X_{-\ell}) = \{(p_\ell, q_\ell) : \sqrt{p_\ell^2+q_\ell^2} > |p_\ell - \ell|,\ q_\ell > 0\}\ . \tag{8.13}$$

Moreover,

$$X_{-\ell} = X_\ell^{-1}\ ,\ \ell \in Z\ . \tag{8.14}$$

In fact, (8.13) and (8.14) follow from the relations

$$X_\ell\, \mathcal{O}_{m,n} = \mathcal{O}_{m-\ell,n+\ell} \tag{8.15}$$

which hold for all $m \geq \ell$ and $n \geq -\ell$.

An explicit construction of the S-matrix $\hat{S}$ can be based on the scattering coefficients $c_\ell^-(p)$ and the mapping X_ℓ. To describe it, it is convenient to write

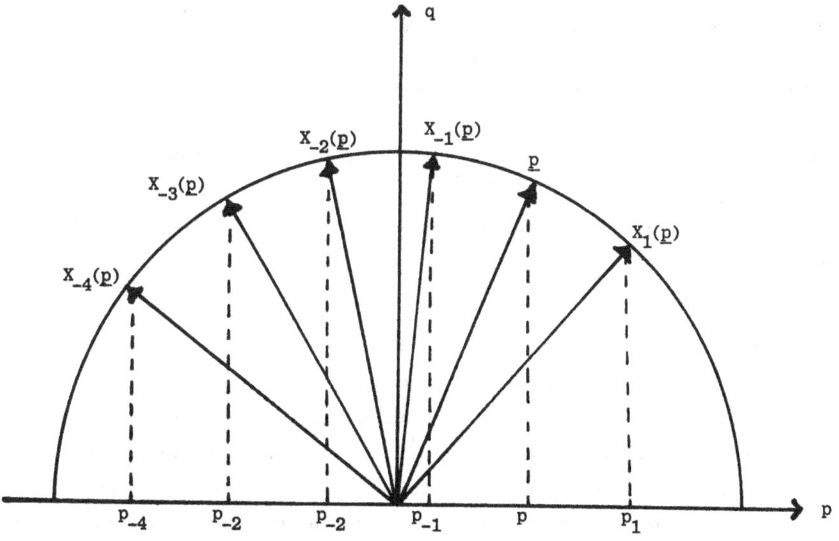

Figure 4. Graphical Construction of the maps X_ℓ.

$$\hat{S} = 1 + \hat{T} \tag{8.16}$$

where 1 denotes the identity operator, and to assume first that $g \in \mathcal{H}_0$ satisfies

$$\text{supp } g \subset \mathcal{O}_{m,n} \; . \tag{8.17}$$

Then the principal result of the section is that

$$\text{supp } \hat{T} \, g \subset \bigcup_{\ell=-m}^{n} \mathcal{O}_{m+\ell,n-\ell} \tag{8.18}$$

and

$$(\hat{T} \, g)(P) = \overline{c_\ell^-(P)} \; g \, (X_\ell(P)) \; , \; P \in \mathcal{O}_{m+\ell,n-\ell} \; . \tag{8.19}$$

This result extends by linearity to general $g \in \mathcal{H}_0$ because functions g with compact supports contained in $\cup_{m,n=0}^{\infty} \mathcal{O}_{m,n}$ are dense in $\mathcal{H}_0$.

It is well known in the theories of scattering by potentials and by bounded obstacles that the S-matrix $\hat{S}$ is a direct integral of a family of unitary operators $\hat{S}(\omega)$ that act on the "energy shell" $p^2 + q^2 = \omega^2$. The analogous property of the S-matrices for diffraction gratings is evident from (8.19) and the properties of the mappings X_ℓ. Note that if $(\omega \cos \theta, \omega \sin \theta) \in R_0^2$ then there is a unique angle $\theta_\ell = \theta_\ell(\omega,\theta)$ such that $0 < \theta_\ell < \pi$ and

$$X_\ell(\omega \cos \theta, \omega \sin \theta) = (\omega \cos \theta_\ell, \omega \sin \theta_\ell) \; . \tag{8.20}$$

With this notation the operator $\hat{S}(\omega) = 1 + \hat{T}(\omega)$ is given by

$$\hat{T}(\omega) \; g(\omega \cos \theta, \omega \sin \theta) = \sum_{\ell=-m}^{n} \overline{c_\ell^-(\omega \cos \theta, \omega \sin \theta)} \; g(\omega \cos \theta_\ell, \omega \sin \theta_\ell) \tag{8.21}$$

provided supp $g \subset \mathcal{O}_{m,n}$. The general case follows by superposition, as before.

§9. Asymptotic Wave Functions for Gratings

The R-B wave expansions (7.5) give the scattered field at time t produced by prescribed sources in the presence of an arbitrary grating that admits no surface waves. In this section it is shown that the author's theory of asymptotic wave functions (Springer Lecture Notes in Mathematics, V. 442, 1975) can be used to simplify the calculation when t is large. The simple case of the reference problem is discussed first.

A state $h \in \mathcal{H}$ may be considered as an initial state for the degenerate grating. The corresponding wave $v_0(t,X)$ is given by (4.10) which is just a Fourier integral. The author's theory is applied in Part 2 to show that

$$(9.1) \qquad v_0(t,X) = e^{-itA_0^{1/2}} h(X) \overset{t\to+\infty}{\sim} v_0^\infty(t,X) = r^{-1/2} F_0(r-t,\theta)$$

where

$$(9.2) \qquad X = (x,y) = (r\cos\theta, r\sin\theta) \ .$$

The wave profile $F_0(\tau,\theta) \in L_2(R \times [0,\pi])$ is defined by the integral

$$(9.3) \qquad F_0(\tau,\theta) = \ell.i.m. \frac{1}{(2\pi)^{1/2}} \int_0^\infty e^{i\tau\omega} \hat{h}_0(\omega\cos\theta,\omega\sin\theta)(-i\omega)^{1/2}\, d\omega \ .$$

The asymptotic equality in (9.1) is meant in the sense of $\mathcal{H}_0$; i.e.,

$$(9.4) \qquad \lim_{t\to+\infty} \| v_0(t,\cdot) - v_0^\infty(t,\cdot)\|_{\mathcal{H}_0} = 0 \ .$$

The structure of F_0 may be made clearer by introducing the function $h_f \in L_2(R^2)$ defined by

$$(9.5) \qquad h_f(X) = \begin{cases} h(X) \ , & y > 0 \ , \\ 0 \ , & y \le 0 \ , \end{cases}$$

and its Fourier transform

$$(9.6) \qquad \hat{h}_f(p,q) = \ell.i.m. \frac{1}{2\pi} \int_{R^2} e^{-i(px+qy)} h_f(x,y)\, dxdy \ .$$

It is clear from (4.1) that

$$(9.7) \qquad \hat{h}_0(p,q) = \hat{h}_f(p,q) + \hat{h}_f(p,-q) \ .$$

Thus if a free wave profile H_f is defined by

$$(9.8) \qquad H_f(\tau,\theta) = \ell.i.m. \frac{1}{(2\pi)^{1/2}} \int_0^\infty e^{i\tau\omega} \hat{h}_f(\omega\cos\theta,\omega\sin\theta)(-i\omega)^{1/2}\, d\omega$$

then (9.3), (9.7) and (9.8) give

$$(9.9) \qquad F_0(\tau,\theta) = F_0^{dir}(\tau,\theta) + F_0^{ref\ell}(\tau,\theta)$$

where

$$(9.10) \qquad \begin{cases} F^{dir}(\tau,\theta) = H_f(\tau,\theta) \ , \\ F^{ref\ell}(\tau,\theta) = H_f(\tau,-\theta) \ . \end{cases}$$

It is clear from (9.1), (9.8), (9.9), (9.10) that F_0^{dir} contains the contribution to v_0^∞ of the momentum components of h that propagate away from the boundary without interaction while $F_0^{ref\ell}$ contains the contributions of momenta that are directed toward the boundary and then reflected.

Now consider the field produced by the <u>same</u> initial state h as in (9.1) in a nondegenerate grating domain G. By the results of §8,

(9.11) $$v(t,\cdot) = e^{-itA^{1/2}} h \overset{t\to+\infty}{\sim} e^{-itA_0^{1/2}} W_+ h \text{ in } \mathcal{H}$$

where

(9.12) $$\Phi_0 W_+ h = \Phi_- h = \hat{h}_- .$$

It follows from (9.1), (9.3), (9.11) and (9.12) that

(9.13) $$v(t,X) = e^{-itA^{1/2}} h(X) \overset{t\to+\infty}{\sim} v^\infty(t,X) = r^{-1/2} F(r-t,\theta)$$

in $\mathcal{H}$ where the wave profile $F(\tau,\theta)$ is given by

(9.14) $$F(\tau,\theta) = \ell.i.m. \frac{1}{(2\pi)^{1/2}} \int_0^\infty e^{i\tau\omega} \hat{h}_-(\omega\cos\theta,\omega\sin\theta)(-i\omega)^{1/2}\, d\omega .$$

Moreover, by (5.1) and (7.1)

(9.15) $$\hat{h}_-(P) = \hat{h}_0(P) + \hat{h}_-^{sc}(P)$$

where

(9.16) $$\hat{h}_-^{sc}(P) = \ell.i.m. \int_G \overline{\psi_-^{sc}(X,P)}\, h(X)\, dX .$$

Comparison of these results with (9.3) and (9.9) gives

(9.17) $$F(\tau,\theta) = F_0^{dir}(\tau,\theta) + F_0^{ref\ell}(\tau,\theta) + F^{sc}(\tau,\theta)$$

where

(9.18) $$F^{sc}(\tau,\theta) = \ell.i.m. \frac{1}{(2\pi)^{1/2}} \int_0^\infty e^{i\tau\omega} \hat{h}_-^{sc}(\omega\cos\theta,\omega\sin\theta)(-i\omega)^{1/2}\, d\omega .$$

The last profile obviously characterizes the deviation of the scattered pulse for G from that for the degenerate grating.

§10. The Scattering of Signals from Remote Sources

The calculation of the profile F^{sc} can be simplified in the case where the wave sources, described by the initial state h, are far from the grating. To make this precise let

$$h(x,y;y_0) = h_0(x,y-y_0) \tag{10.1}$$

where $h_0 \in \mathcal{H}$ is a fixed function. Under this hypothesis it is shown in Part 2 that

$$F^{sc}(\tau,\theta) = \frac{1}{(2\pi)^{1/2}} \int_0^\infty e^{i\tau\omega}\, \hat{T}(\omega)\, \hat{F}^{ref\ell}(\omega,\theta)\, d\omega + o(1) \tag{10.2}$$

where $\hat{F}^{ref\ell}(\omega,\theta) = \hat{h}_f(\omega \cos\theta, -\omega \sin\theta, y_0)$ is the Fourier transform, with respect to τ, of $F^{ref\ell}(\tau,\theta)$ and the error term $o(1)$ tends to zero in $L_2(R \times [0,\pi])$ when $y_0 \to \infty$. Thus apart from this error the scattered wave profile is determined by the T-matrix acting on the reflected part of the incident wave profile.

A case of special interest arises when

$$\operatorname{supp} F_0^{ref\ell} = K \subset \mathcal{O}_{m,n} \tag{10.3}$$

for some m and n. This defines an incident asymptotic wave profile $F_0^{ref\ell}$ that might be called a narrow beam. Write

$$K_\ell = X_\ell(K) \subset \mathcal{O}_{m-\ell,n+\ell} \tag{10.4}$$

for $-m \le \ell \le n$ and let

$$\Gamma_\ell = \{(p,q) = (\omega \cos\theta, \omega \sin\theta) : \omega > 0,\ \alpha_\ell \le \theta \le \beta_\ell\} \tag{10.5}$$

be the smallest sector such that

$$K_\ell \subset \Gamma_\ell\ ,\quad -m \le \ell \le n\ . \tag{10.6}$$

It is easy to verify that the sectors are disjoint. Moreover, (8.21) and (10.2) imply that, apart from the error $o(1)$, the support of $F^{sc}(\tau,\theta)$ lies in $\cup_{\ell=-m}^{n} \Gamma_\ell$. Thus for a narrow beam satisfying (10.3) the scattered pulse is concentrated, apart from an error $o(1)$, in the $m+n+1$ sectors Γ_ℓ associated with K.

Part 2
Mathematical Theory

The purpose of this Part 2 is to penetrate more deeply into the theory described in Part 1 and to develop the results in a precise form with complete hypotheses and full mathematical proofs. The work is based squarely on functional analysis. The reader should have good knowledge of the theory of unbounded selfadjoint operators in Hilbert spaces, as developed in Dunford and Schwartz [9] or any of the many other good texts. Other prerequisites include the theory of Sobolev spaces and the L_2 theory of elliptic boundary value problems, as presented in the texts of S. Agmon [1] or Lions and Magenes [14], simple facts from the theory of Fréchet spaces and distribution theory, and the elements of the abstract theory of Riemann surfaces (Narasimhan [16]).

No attempt is made in Part 2 to justify explicitly all the statements made in Part 1. It was in the spirit of Part 1 that minor, or highly technical, hypotheses were omitted. Nevertheless, every result mentioned in Part 1 has a more precise counterpart in Part 2. The notation of Part 2 is consistent with that of Part 1 but is frequently more elaborate.

The class of grating domains G admitted in Part 2 is very general. In fact, for the Dirichlet boundary condition G may be an <u>arbitrary</u> grating domain; i.e., an open connected subset $G \subset R^2$ that satisfies properties (2.2) and (2.3) of Part 1. For the Neumann case a mild regularity condition is imposed. In neither case need ∂G be given by a function $y = h(x)$ nor even be a unicursal curve.

The principal technical difficulties of Part 2 occur in sections 4, 6 and 8. For this reason most of the proofs of the results of these chapters are collected in the technical sections 5, 7 and 9 which may be omitted in a first reading.

§1. Grating Domains and Grating Propagators

The plane diffraction gratings that are studied in this monograph are the boundaries of the class of planar domains G defined by the following properties.

(1.1) G is contained in a half-plane.

(1.2) G contains a smaller half-plane.

(1.3) G is invariant under translation through a distance $a > 0$.

Domains with these properties will be called grating domains. The half-plane of (1.2) is necessarily parallel to that of (1.1) and the translation of (1.3) is necessarily parallel to the edges of these half-planes. The smallest $a > 0$ for which (1.3) holds is called the primitive grating period. It exists for all gratings except the degenerate grating for which G is a half-plane.

It will be convenient to introduce Cartesian coordinates

$$X = (x,y) \in R^2 \tag{1.4}$$

in the plane of G such that the x-axis is parallel to the edges of the half-planes of (1.1), (1.2) and to identify G with the corresponding domain (open connected set) $G \subset R^2$. With this convention if

$$R_c^2 = \{X \in R^2 \mid y > c\} \tag{1.5}$$

then, for a suitable orientation of the coordinate axes, conditions (1.1), (1.2) can be written

$$R_h^2 \subset G \subset R_0^2 \text{ for some } h > 0 \tag{1.6}$$

and the translation invariance (1.3) takes the form

$$G + (a,0) = G \tag{1.7}$$

where $a > 0$ is the primitive period of G.

The eigenfunction expansion theory for R-B waves that satisfy the Dirichlet boundary condition is developed below for arbitrary grating domains. For R-B waves that satisfy the Neumann boundary condition the following additional conditions are imposed on ∂G, the frontier of G.

(1.8) G has the local compactness property, and

(1.9) there exists an $x_0 \in R$ such that the set $\partial G \cap \{(x_0,y) \mid y \geq 0\}$ is finite and each (x_0,y) in the set has a neighborhood in R^2 in which ∂G is a regular curve of class C^3.

Condition (1.8) was introduced in [30] where it was denoted by $G \in LC$. It is a mild regularity property of ∂G. A simple sufficient condition for $G \in LC$ is the "finite tiling condition" of [30, p. 63]. Grating domains that satisfy (1.8) and (1.9) will be said to have property S, written $G \in S$. The class includes all the piece-wise smooth gratings that arise in applications. Examples include the domains $G = \{X \mid y > h(x)\}$ where $h(x)$ is bounded, piece-wise smooth and has period a. A special case is DeSanto's comb grating for which

$$h(x) = \begin{cases} h > 0 \text{ for } x = 0 \\ 0 \text{ for } -\frac{a}{2} \leq x < 0 \text{ and } 0 < x \leq \frac{a}{2} . \end{cases} \tag{1.10}$$

The Hilbert space theory of solutions of the wave equation in arbitrary domains $G \subset R^n$, developed by the author in [28,30], provides the foundation for the analysis of scattering by diffraction gratings given below. The basic Hilbert space of the theory is the Lebesgue space $L_2(G)$ with scalar product

$$(u,v) = \int_G \overline{u(X)}\, v(X)\, dX . \tag{1.11}$$

In addition, the definition of the grating propagators makes use of the Sobolev spaces

$$L_2^m(G) = L_2(G) \cap \{u \mid D_1^{\alpha_1} D_2^{\alpha_2} u \in L_2(G) \text{ for } \alpha_1 + \alpha_2 \leq m\} , \tag{1.12}$$

where $D_1 = \partial/\partial x$, $D_2 = \partial/\partial y$ and m is a positive integer, and the space

$$L_2^1(\Delta,G) = L_2^1(G) \cap \{u \mid \Delta u \in L_2(G)\} \tag{1.13}$$

where $\Delta = D_1^2 + D_2^2$ is the Laplacian in R^2. In these definitions the differential operators are to be interpreted in the distribution-theoretic sense (cf. [28,30]).

The grating propagators for a grating domain G are selfadjoint realizations in $L_2(G)$ of $-\Delta$, acting on sets of functions that satisfy the Neumann or Dirichlet boundary conditions. These operators will be denoted by $A^N(G)$

and $A^D(G)$, respectively. Their domains are subsets of $L_2^1(\Delta,G)$ that satisfy the boundary conditions in a form appropriate to arbitrary domains G. In particular, functions $u \in D(A^N(G))$ are required to satisfy the generalized Neumann condition

$$\int_G \{(\Delta u)\overline{v} + \nabla u \cdot \nabla\overline{v}\}\, dX = 0 \tag{1.14}$$

for all $v \in L_2^1(G)$. In fact, if one defines

$$L_2^N(\Delta,G) = L_2^1(\Delta,G) \cap \{u \mid (1.14) \text{ holds for all } v \in L_2^1(G)\} , \tag{1.15}$$

$D(A^N(G)) = L_2^N(\Delta,G)$ and $A^N(G)u = -\Delta u$ then $A^N(G)$ is a selfadjoint non-negative operator in $L_2(G)$. This characterization was proved in [30]. It may also be derived from T. Kato's theory of sesquilinear forms in Hilbert space [13, Ch. 6]. It is known that if ∂G is a smooth curve then $D(A^N(G)) \subset L_2^2(G)$ and ∇u has a trace in $L_2(\partial G)$ which satisfies the Neumann boundary condition [30].

To define the grating propagator $A^D(G)$ associated with the Dirichlet boundary condition let

$$L_2^D(G) = \text{closure of } C_0^\infty(G) \text{ in } L_2^1(G) \tag{1.16}$$

and define

$$L_2^D(\Delta,G) = L_2^D(G) \cap L_2^1(\Delta,G) , \tag{1.17}$$

$D(A^D(G)) = L_2^D(\Delta,G)$ and $A^D(G)u = -\Delta u$. Then Kato's theory of sesquilinear forms may be used to show that $A^D(G)$ is also a selfadjoint non-negative operator in $L_2(G)$. Moreover, it is known that if ∂G is a smooth curve then every $u \in L_2^1(G)$ has a trace $u|\partial G \in L_2(\partial G)$ and every $u \in L_2^D(G)$ satisfies $u|\partial G = 0$ [14].

The grating propagators $A^N(G)$ and $A^D(G)$ will be shown to have pure continuous spectra. It follows that the R-B wave eigenfunctions must be generalized eigenfunctions which are not in $L_2(G)$. To define them it will be convenient to define extensions of $A^N(G)$ and $A^D(G)$ which act in the space

$$L_2^{\ell oc}(G) = \mathcal{D}'(G) \cap \{u \mid u \in L_2(K \cap G) \text{ for all compact } K \subset R^2\} \tag{1.18}$$

where $\mathcal{D}'(G)$ is the set of all distributions on G. The following subsets of $L_2^{\ell oc}(G)$ are also needed

$$L_2^{m,\ell oc}(G) = L_2^{\ell oc}(G) \cap \{u \mid D_1^{\alpha_1}D_2^{\alpha_2}u \in L_2^{\ell oc}(G) \text{ for } \alpha_1 + \alpha_2 \le m\} , \tag{1.19}$$

$$(1.20) \qquad L_2^{1,\ell oc}(\Delta,G) = L_2^{1,\ell oc}(G) \cap \{u \mid \Delta u \in L_2^{\ell oc}(G)\} .$$

These linear spaces are all Fréchet spaces (locally convex topological vector spaces which are metrizable and complete [9]) under suitable definitions of the topologies. Thus $L_2^{\ell oc}(G)$ is a Fréchet space with family of semi-norms

$$(1.21) \qquad \rho_K(u) = \left(\int_{K\cap G} |u(X)|^2 \, dX\right)^{1/2}$$

indexed by the compact sets $K \subset R^2$. Similarly, $L_2^{m,\ell oc}(G)$ is a Fréchet space with family of semi-norms

$$(1.22) \qquad \rho_K(u) = \left(\int_{K\cap G} \sum_{\alpha_1+\alpha_2\leq m} |D_1^{\alpha_1}D_2^{\alpha_2}u(X)|^2 \, dX\right)^{1/2}$$

and $L_2^{1,\ell oc}(\Delta,G)$ is a Fréchet space with family of semi-norms

$$(1.23) \qquad \rho_k(u) = \left(\int_{K\cap G} \{|u(X)|^2 + |\nabla u(X)|^2 + |\Delta u(X)|^2\} \, dX\right)^{1/2} .$$

The following additional notation is used below:

$$(1.24) \qquad L_2^{com}(G) = L_2(G) \cap \mathcal{E}'(R^2)$$

$$(1.25) \qquad L_2^{1,com}(G) = L_2^1(G) \cap L_2^{com}(G)$$

where $\mathcal{E}'(R^2)$ denotes the set of all distributions on R^2 with compact supports.

The local grating propagator $A^{N,\ell oc}(G)$ for G and the Neumann boundary condition is the extension of $A^N(G)$ in $L_2^{\ell oc}(G)$ defined by

$$D(A^{N,\ell oc}(G)) = L_2^{N,\ell oc}(\Delta,G)$$

(1.26)

$$\equiv L_2^{1,\ell oc}(\Delta,G) \cap \{u \mid (1.14) \text{ holds for all } v \in L_2^{1,com}(G)\}$$

and

$$(1.27) \qquad A^{N,\ell oc}(G)\, u = -\Delta u \text{ for all } u \in D(A^{N,\ell oc}(G)) .$$

Similarly, the local grating propagator $A^{D,\ell oc}(G)$ for G and the Dirichlet boundary condition is the extension of $A^D(G)$ in $L_2^{\ell oc}(G)$ defined by

$$(1.28) \qquad D(A^{D,\ell oc}(G)) = L_2^{D,\ell oc}(\Delta,G) \equiv L_2^{D,\ell oc}(G) \cap L_2^{1,\ell oc}(\Delta,G)$$

where

(1.29) $$L_2^{D,\ell oc}(G) = \text{Closure of } C_0^\infty(G) \text{ in } L_2^{1,\ell oc}(G)$$

and

(1.30) $$A^{D,\ell oc}(G)u = -\Delta u \text{ for all } u \in D(A^{D,\ell oc}(G)) .$$

The spectral analysis and eigenfunction expansions for $A^N(G)$ and $A^D(G)$ are nearly identical. To emphasize this, and to simplify the notation, the symbol A will be used to denote either $A^N(G)$ or $A^D(G)$ in stating results that are valid for both. Similarly, the symbol $A^{\ell oc}$ will denote $A^{N,\ell oc}(G)$ or $A^{D,\ell oc}(G)$ except where a distinction is necessary.

The spectral theory of $A^N(G)$ and $A^D(G)$ will be developed by perturbation theory, beginning with the degenerate grating R_0^2. The grating propagators for this case will be denoted by

(1.31) $$A_0^N = A^N(R_0^2) , \qquad A_0^D = A^D(R_0^2)$$

and

(1.32) $$A_0^{N,\ell oc} = A^{N,\ell oc}(R_0^2) , \qquad A_0^{D,\ell oc} = A^{D,\ell oc}(R_0^2)$$

and the condensed notation A_0 for A_0^N or A_0^D and $A_0^{\ell oc}$ for $A_0^{N,\ell oc}$ or $A_0^{D,\ell oc}$ will be used.

The spectral analysis of A_0 can be carried out by separation of variables and is essentially elementary. Thus D_1^2 is essentially selfadjoint in $L_2(R)$ with complete family of generalized eigenfunctions $\{(2\pi)^{-1/2} \exp(ipx) \mid p \in R\}$. Similarly, D_2^2 and the Neumann boundary condition define a selfadjoint operator in $L_2(0,\infty)$ with complete family $\{(2/\pi)^{1/2} \cos q\, y \mid q > 0\}$, while D_2^2 and the Dirichlet boundary condition define a second selfadjoint operator in $L_2(0,\infty)$ with complete family $\{(2/\pi)^{1/2} \sin q\, y \mid q > 0\}$. It follows that the products

(1.33) $$\psi_0^N(X,p,q) = \frac{1}{\pi} e^{ip\cdot x} \cos q\, y , \quad (p,q) \in R_0^2$$

(1.34) $$\psi_0^D(X,p,q) = \frac{1}{\pi} e^{ip\cdot x} \sin q\, y , \quad (p,q) \in R_0^2$$

are in $D(A_0^{N,\ell oc})$ and $D(A_0^{D,\ell oc})$, respectively, and define complete families of generalized eigenfunctions for A_0^N and A_0^D. More precisely, if ψ_0 is used to denote either ψ_0^N or ψ_0^D then the classical Plancherel theory can be used to derive an eigenfunction expansion and spectral decomposition for

(1.35) $$A_0 = \int_0^\infty \mu \, d\Pi_0(\mu)$$

which may be formulated as follows. First, for all $f \in L_2(R_0^2)$ the limit

(1.36) $$\hat{f}_0(p,q) = \underset{M\to\infty}{\ell.i.m.} \int_0^M \int_{-M}^M \overline{\psi_0(X,p,q)} \, f(X) \, dX$$

exists in $L_2(R_0^2)$,

(1.37) $$f(X) = \underset{M\to\infty}{\ell.i.m.} \int_0^M \int_{-M}^M \psi_0(X,p,q) \, \hat{f}_0(p,q) \, dpdq$$

in $L_2(R_0^2)$, and

(1.38) $$\|f\|_{L_2(R_0^2)} = \|\hat{f}_0\|_{L_2(R_0^2)} .$$

Moreover, the spectral family of A_0 is given by

(1.39) $$\Pi_0(\mu) \, f(X) = \int_{\{(p,q)\,|\,p^2+q^2\leq\mu,q>0\}} \psi_0(X,p,q) \, \hat{f}_0(p,q) \, dpdq .$$

Finally, if a linear operator $\Phi_0 : L_2(R_0^2) \to L_2(R_0^2)$ is defined by $\Phi_0 f = \hat{f}_0$ then Φ_0 is unitary.

The principal result of this monograph is a generalization of this eigenfunction expansion and spectral analysis that is valid for the operator $A^D(G)$ in arbitrary grating domains G and for the operator $A^N(G)$ in grating domains $G \in S$. In these generalizations the R-B waves play the role of the eigenfunctions ψ_0.

§2. Rayleigh-Bloch Waves

It will be assumed in the remainder of the report that the unit of length has been chosen to make the grating period $a = 2\pi$. This normalization, which simplifies many of the equations, does not limit the generality of the theory because the general case can be obtained by a simple change of units.

The definition of the R-B waves can be motivated by considering the reflection by a grating of a plane wave

(2.1) $$\psi^{inc}(X,p,q) = (2\pi)^{-1} \exp\{i(px - qy)\}, \quad (p,q) \in R_0^2 .$$

Note that the effect of translating ψ^{inc} by the grating period 2π is to multiply it by a factor of modulus 1:

(2.2) $$\psi^{inc}(x + 2\pi, y, p, q) = \exp\{2\pi ip\}\, \psi^{inc}(x, y, p, q) .$$

Since G is invariant under this translation the reflected wave, if it is uniquely determined by ψ^{inc}, must also have property (2.2). This suggests the

Definition. A function $\psi \in L_2^{1,\ell oc}(\Delta, G)$ is said to be an R-B wave for G if and only if there exist numbers $p \in R$ and $\omega \geq 0$ such that

(2.3) $$\psi(x + 2\pi, y) = \exp\{2\pi ip\}\, \psi(x,y) \text{ in } G ,$$

(2.4) $$\Delta\psi + \omega^2\psi = 0 \text{ in } G, \text{ and}$$

(2.5) $$\psi(X) \text{ is bounded in } G.$$

If, in addition,

(2.6) $$\psi \in D(A^{\ell oc})$$

then ψ is said to be an R-B wave for A.

The parameters ω and p will be called the frequency and x-momentum of the R-B wave, respectively. Note that p is only determined modulo 1 by (2.3). The x-momentum that satisfies

(2.7) $$-\frac{1}{2} < p \leq \frac{1}{2}$$

will be called the reduced x-momentum of ψ. Property (2.3) is sometimes called quasi-periodicity or p-periodicity. It is equivalent to the property that

(2.8) $$\psi(x,y) = \exp\{ipx\}\, \phi(x,y) \text{ for all } (x,y) \in G$$

where

(2.9) $$\phi(x + 2\pi, y) = \phi(x,y) \text{ for all } (x,y) \in G .$$

Solutions of the Helmholtz equation (2.4) are known to be analytic functions. In particular, each R-B wave for A satisfies $\psi \in C^\infty(G)$. Hence, the function ϕ in (2.8) is in $C^\infty(G)$ and has period 2π in x since $R_h^2 \subset G$. It follows from classical convergence theory for Fourier series that ψ has an expansion

(2.10) $$\psi(x,y) = \sum_{\ell \in Z} \psi_\ell(y) \exp\{i(p + \ell)x\} , \quad (x,y) \in R_h^2 ,$$

where Z denotes the set of all integers. The series converges absolutely and uniformly on compact subsets of R_h^2. Moreover, the partial derivatives of ψ have expansions of the same form which may be calculated from (2.10) by term-by-term differentiation and which have the same convergence properties. It follows that the coefficients $\psi_\ell(y)$ in (2.10) must satisfy

$$\psi_\ell''(y) + (\omega^2 - (p + \ell)^2)\, \psi_\ell(y) = 0 \text{ for } y > h\ . \tag{2.11}$$

Hence the terms in the expansion (2.10) have the following forms, depending on the relative magnitudes of ω and $|p + \ell|$.

$\underline{\omega > |p + \ell|}$. In this case there exist constants c_ℓ^+ and c_ℓ^- such that

$$\psi_\ell(y) \exp\{i(p+\ell)x\} = c_\ell^+ \exp\{i(p_\ell x + q_\ell y)\} + c_\ell^- \exp\{i(p_\ell x - q_\ell y)\} \tag{2.12}$$

where

$$p_\ell = p + \ell\ , \quad q_\ell = (\omega^2 - (p + \ell)^2)^{1/2} > 0\ . \tag{2.13}$$

The two terms in (2.12) describe plane waves propagating in the directions $(p_\ell, \pm q_\ell)$. Since $p_\ell^2 + q_\ell^2 = \omega^2$ these vectors lie on the circle of radius ω with center at the origin and their x-components differ by integers. Clearly there are only finitely many such terms.

$\underline{\omega < |p + \ell|}$. In this case $\psi_\ell(y)$ is a linear combination of real exponentials in y and the boundedness condition (2.5) implies that

$$\psi_\ell(y) \exp\{i(p+\ell)x\} = c_\ell \exp\{-((p+\ell)^2 - \omega^2)^{1/2} y\} \exp\{i(p+\ell)x\} \tag{2.14}$$

where $((p+\ell)^2 - \omega^2)^{1/2} > 0$. In the application to diffraction gratings terms of this type will be interpreted as surface waves.

$\underline{\omega = |p + \ell|}$. In this limiting case $\psi_\ell(y)$ is a linear combination of 1 and y and (2.5) implies that

$$\psi_\ell(y) \exp\{i(p + \ell)x\} = c_\ell \exp\{i(p + \ell)x\}\ . \tag{2.15}$$

Physically, (2.15) describes a plane wave that propagates parallel to the grating; i.e., a grazing wave. These waves divide the plane waves (2.12) from the surface waves (2.14). The frequencies $\{\omega = |p + \ell| \mid \ell \in Z\}$ are called the cut-off frequencies for R-B waves with x-momentum p.

An R-B wave ψ for G (for A) which satisfies the additional conditions

$$c_\ell^- = 0 \text{ (resp. } c_\ell^+ = 0) \text{ for all } \ell \text{ such that } \omega > |p + \ell| \tag{2.16}$$

will be said to be an outgoing (resp., incoming) R-B wave for G (for A). If

$$c_\ell^- = c_\ell^+ = 0 \text{ for all } \ell \text{ such that } \omega > |p + \ell| \tag{2.17}$$

then ψ will be said to be an R-B surface wave for G (for A). Of course an R-B surface wave for A is both an outgoing and an incoming R-B wave for A. It is interesting that these are the only outgoing or incoming R-B waves for A. This is a consequence of

Theorem 2.1. Every outgoing (resp., incoming) R-B wave for A is an R-B surface wave for A.

A proof of this result has been given by Alber [3] in the case where ∂G is a curve of class C^2. The method is to apply Green's theorem to the R-B wave ψ for A and its conjugate in the region $G \cap \{X \mid -\pi < x < \pi, y < R\}$. In the case of an outgoing R-B wave for A this yields the equation

$$\sum_{\omega > |p+\ell|} (\omega^2 - (p + \ell)^2)^{1/2} \; |c_\ell^+|^2 = 0 \tag{2.18}$$

which implies that $c_\ell^+ = 0$ when $\omega > |p + \ell|$. For general grating domains the application of Green's theorem must be based on the generalized boundary conditions, as in [30, p. 57].

It will be seen in §4 that diffraction gratings may indeed support R-B surface waves and the question arises whether geometric criteria for the non-existence of such waves can be found. In the case of the Dirichlet boundary condition such a criterion was found by Alber [3] by adapting a method of F. Rellich [20] and D. M. Eidus [11]. Specialized to the grating domains considered here, Alber's theorem implies

Theorem 2.2. Let

$$G = \{X \mid y > h(x) \text{ for all } x \in R\} \tag{2.19}$$

where $h \in C^2(R)$ and $h(x + 2\pi) = h(x)$ for all $x \in R$. Then $A^D(G)$ has no R-B surface waves.

Theorem 2.1 implies that R-B waves for A may be determined, modulo R-B surface waves, by specifying either the coefficients c_ℓ^- with $\omega > |p + \ell|$ (the incoming plane waves) or the coefficients c_ℓ^+ with $\omega > |p + \ell|$ (the outgoing plane waves). R-B waves for A that contain a single incoming or outgoing plane wave will be used in the R-B wave expansions given in §8 below. These are the grating waves originally introduced by Rayleigh. Physically, they are the wave fields produced when the grating is illuminated by a single plane wave. Here they will be called R-B diffracted plane wave eigenfunctions for A or, for brevity, R-B wave eigenfunctions for A. There are two families determined by the presence of a single incoming or outgoing plane wave, respectively. The plane waves $\psi^{inc}(X,p,q)$ and $\psi^{inc}(X,p,-q)$ defined by (2.1) are incoming and outgoing R-B waves, respectively, with x-momentum p and frequency

$$\omega = \omega(p,q) = (p^2 + q^2)^{1/2} \quad . \tag{2.20}$$

The scattering of these waves by a grating will produce outgoing (resp., incoming) R-B waves with the same x-momentum and frequency. Hence the R-B wave eigenfunctions may be defined as follows.

Definition. An outgoing R-B diffracted plane wave for A with momentum $(p,q) \in R_0^2$ is a function $\psi_+(X,p,q)$ such that

$$\psi_+(\cdot,p,q) \text{ is an R-B wave for A, and} \tag{2.21}$$

$$\psi_+(X,p,q) = \psi^{inc}(X,p,q) + \psi_+^{sc}(X,p,q) \tag{2.22}$$

where ψ_+^{sc} is an outgoing R-B wave for G. Similarly, an incoming R-B diffracted plane wave for A with momentum $(p,q) \in R_0^2$ is a function $\psi_-(X,p,q)$ such that

$$\psi_-(\cdot,p,q) \text{ is an R-B wave for A, and} \tag{2.23}$$

$$\psi_-(X,P,q) = \psi^{inc}(X,p,-q) + \psi_-^{sc}(X,p,q) \tag{2.24}$$

where ψ_-^{sc} is an incoming R-B wave for G.

The uniqueness of $\psi_\pm(X,p,q)$ modulo R-B surface waves follows from Theorem 2.1, as was remarked above. Their existence for the class of gratings defined in §1 is proved in §8 below. Note also that the defining properties imply that

$$\psi_-(X,p,q) = \overline{\psi_+(X,-p,q)} \ . \tag{2.25}$$

Hence the existence of the family ψ_- follows from that of ψ_+.

In the half-plane R_h^2 above the grating the R-B waves $\psi_\pm$ have Fourier expansions (2.10). For the function ψ_+ the expansion has the form

$$\begin{aligned} \psi_+(x,y,p,q) &= (2\pi)^{-1} \exp\{i(px - qy)\} \\ &+ (2\pi)^{-1} \sum_{(p+\ell)^2<p^2+q^2} c_\ell^+(p,q) \exp\{i(p_\ell x + q_\ell y)\} \\ &+ (2\pi)^{-1} \sum_{(p+\ell)^2\ge p^2+q^2} c_\ell^+(p,q) \exp\{ip_\ell x\} \exp\{-((p+\ell)^2 - p^2 - q^2)^{1/2}\, y\} \end{aligned} \tag{2.26}$$

where

$$(p_\ell, q_\ell) = (p + \ell, \{p^2 + q^2 - (p + \ell)^2\}^{1/2}) \in R_0^2 \tag{2.27}$$

defines the momentum of the reflected plane wave of order ℓ. Similarly,

$$\psi_-(x,y,p,q) = (2\pi)^{-1} \exp \{i(px + qy)\}$$

$$+ (2\pi)^{-1} \sum_{(p+\ell)^2<p^2+q^2} c_\ell^-(p,q) \exp \{i(p_\ell x - q_\ell y)\} \tag{2.28}$$

$$+ (2\pi)^{-1} \sum_{(p+\ell)^2\geq p^2+q^2} c_\ell^-(p,q) \exp \{ip_\ell x\} \exp \{-((p+\ell)^2 - p^2 - q^2)^{1/2} y\}.$$

The relation (2.25) implies that the coefficients $c_\ell^\pm(p,q)$ in (2.26), (2.28) satisfy

$$c_\ell^-(p,q) = \overline{c_{-\ell}^+(-p,q)} \quad \text{for all } (p,q) \in R_0^2 \text{ and } \ell \in Z . \tag{2.29}$$

The surface wave terms in (2.26) and (2.28) are exponentially decreasing functions of y except when the wave frequency $\omega(p,q) = (p^2 + q^2)^{1/2} = |p + \ell|$ for some $\ell \in Z$. These are precisely the cut-off frequencies mentioned above. In momentum space they form the exceptional set

$$E = R_0^2 \cap \bigcup_{\ell\in Z} \{(p,q) \mid \sqrt{p^2 + q^2} = |p + \ell|\} \tag{2.30}$$

E is a set of confocal parabolas with foci at (0,0), axes along the p-axis and directrices $p + \ell = 0$, $\ell \in Z$. Two members of the family with directrices $p + \ell = 0$, $p + m = 0$ are disjoint if ℓ and m have the same sign and intersect orthogonally if ℓ and m have opposite signs. The family E thus divides R_0^2 into a system of curvilinear rectangles.

In the special case of the degenerate grating R_0^2 comparison of (1.33), (1.34) with (2.26), (2.28) shows that for the Neumann case $\psi_{0\pm}^N = \psi_0^N$, $c_0^\pm(p,q) = 1$ and all other $c_\ell^\pm(p,q) = 0$. Similarly, for the Dirichlet case $\psi_{0\pm}^D = \mp i\psi_0^D$, $c_0^\pm(p,q) = -1$ and all others $c_\ell^\pm(p,q) = 0$. Thus in these cases there is no scattering into higher order grating modes or surface waves, as was to be expected. Note that the defining properties (2.22), (2.24) can be rewritten as

$$\psi_\pm(X,p,q) = \psi_{0\pm}(X,p,q) + \psi_\pm'(X,p,q) \tag{2.31}$$

where $\psi_{0\pm}$ is defined as above and ψ_+' and ψ_-' are, respectively, outgoing and incoming R-B waves for G. This decomposition exhibits the R-B wave eigenfunctions for G as perturbations of those for R_0^2. The decomposition is used below for the construction of $\psi_\pm$ and the derivation of the eigenfunction expansions.

§3. The Reduced Grating Propagator A_p.

The quasi-periodicity property (2.3) of the R-B waves implies that they are completely determined by their values in the domain

$$\Omega = G \cap \{X \mid -\pi < x < \pi\} . \tag{3.1}$$

Moreover, (2.3) and the equation obtained from it by x-differentiation define boundary conditions that must be satisfied by R-B waves on the portions of $\partial\Omega$ where $x = \pm\pi$. These observations are used below to show that the R-B surface waves and diffracted plane waves for G are eigenfunctions and generalized eigenfunctions, respectively, of a p-dependent selfadjoint realization of $-\Delta$ in $L_2(\Omega)$. This operator, which will be denoted by A_p and called the reduced grating propagator, provides a basis for the construction of the R-B waves for G.

The definition of the grating domains in §1 implies that the reduced grating domains Ω satisfy

$$B_h \subset \Omega \subset B_0 \text{ for some } h > 0 \tag{3.2}$$

where

$$B_c = R_c^2 \cap \{X \mid -\pi < x < \pi\} = \{X \mid -\pi < x < \pi,\ y > c\} . \tag{3.3}$$

The notation

$$\gamma = \{y \mid (\pi,y) \in G\} = \{y \mid (-\pi,y) \in G\} \tag{3.4}$$

will also be used. The definition of the reduced grating propagators $A_p^N(\Omega)$ and $A_p^D(\Omega)$ associated with Ω and the two boundary conditions will be based on the function space

$$L_2^{1,p}(\Omega) = L_2^1(\Omega) \cap \{u \mid u(\pi,y) = \exp\{2\pi i p\}\, u(-\pi,y),\ y \in \gamma\} \tag{3.5}$$

Sobolev's imbedding theorem [1] implies that every $u \in L_2^1(\Omega)$ has boundary values $u(\pm\pi,y)$ in $L_2^{loc}(\gamma)$ and $L_2^{1,p}(\Omega)$ is a closed subspace of $L_2^1(\Omega)$.

The operator $A_p^N(\Omega)$ is defined by

$$D(A_p^N(\Omega)) = L_2^{1,p}(\Omega) \cap L_2^1(\Delta,\Omega) \cap \{u \mid \int_\Omega \{(\Delta u)\bar{v} + \nabla u \cdot \nabla\bar{v}\}\, dX = 0 \text{ for } v \in L_2^{1,p}(\Omega)\}$$

(3.6)

and $A_p^N(\Omega)u = -\Delta u$. It can be shown that $A_p^N(\Omega)$ is the selfadjoint non-negative operator in $L_2(\Omega)$ associated via Kato's theory with the sesquilinear form defined by the Dirichlet integral acting on the domain $L_2^{1,p}(\Omega)$. By applying elliptic regularity theory [1] and Sobolev's imbedding theorem it can also be shown that every $u \in D(A_p^N(\Omega))$ satisfies the p-periodic boundary conditions

$$(3.7)\qquad \begin{cases} u(\pi,y) = \exp\{2\pi ip\}\, u(-\pi,y)\ , & y \in \gamma \\ D_1u(\pi,y) = \exp\{2\pi ip\}\, D_1u(-\pi,y)\ , & y \in \gamma\ . \end{cases}$$

Moreover, if ∂G is a smooth curve then it follows from (1.14) as in §1 that functions $u \in D(A_p^N(\Omega))$ satisfy the Neumann boundary condition on

$$(3.8)\qquad \Gamma = \partial G \cap \overline{\Omega}$$

where $\overline{\Omega}$ is the closure of Ω in R^2.

To define $A_p^D(\Omega)$ several additional function spaces are needed. The subset of $C^\infty(G)$ consisting of functions that satisfy

$$(3.9)\qquad \phi(x+2\pi,y) = \exp\{2\pi ip\}\,\phi(x,y) \text{ for all } (x,y) \in G$$

$$(3.10)\qquad \text{supp}\,\phi \subset G \cap \{X \mid y < \rho\} \text{ where } \rho = \rho(\phi)\ , \text{ and}$$

$$(3.11)\qquad \text{dist}\,(\text{supp}\,\phi, \partial G) > 0\ ,$$

will be denoted by $C_p^\infty(G)$:

$$(3.12)\qquad C_p^\infty(G) = C^\infty(G) \cap \{\phi \mid (3.9),\ (3.10) \text{ and } (3.11) \text{ hold}\}\ .$$

The restrictions of such functions to Ω defines

$$(3.13)\qquad C_p^\infty(\Omega) = \{\psi = \phi|_\Omega \mid \phi \in C_p^\infty(G)\}\ .$$

Finally,

$$(3.14)\qquad L_2^{D,p}(\Omega) = \text{Closure in } L_2^1(\Omega) \text{ of } C_p^\infty(\Omega)\ .$$

The operator $A_p^D(\Omega)$ is defined by

$$D(A_p^D(\Omega)) = L_2^{D,p}(\Omega) \cap L_2^1(\Delta,\Omega) \cap \{u \mid \int_\Omega \{(\Delta u)\overline{v} + \nabla u \cdot \nabla\overline{v}\}\, dX = 0 \text{ for } v \in L_2^{D,p}(\Omega)\}$$

(3.15)

and $A_p^D(\Omega)u = -\Delta u$. In this case it can be shown that $A_p^D(\Omega)$ is the selfadjoint non-negative operator associated via Kato's theory with the sesquilinear form defined by the Dirichlet integral acting on the domain $L_2^{D,p}(\Omega)$. Again, functions in $D(A_p^D(\Omega))$ satisfy the p-periodic boundary conditions (3.7). Moreover, if ∂G is a smooth curve then functions $u \in D(A_p^D(\Omega))$ satisfy the Dirichlet boundary condition on Γ.

Each of the operators $A_p^N(\Omega)$ and $A_p^D(\Omega)$ will be shown in §6 to have a continuous spectrum plus possible point spectrum. To define corresponding generalized eigenfunctions it will be convenient to define extensions of $A_p^N(\Omega)$ and $A_p^D(\Omega)$ in $L_2^{\ell oc}(\Omega)$. The following subsets of $L_2^{\ell oc}(\Omega)$ are also needed:

$$(3.16) \qquad L_2^{1,p,\ell oc}(\Omega) = L_2^{1,\ell oc}(\Omega) \cap \{u \mid u(\pi,y) = \exp\{2\pi i p\}\, u(-\pi,y),\ y \in \gamma\}\ ,$$

$$(3.17) \qquad L_2^{D,p,\ell oc}(\Omega) = \text{Closure of } C_p^\infty(\Omega) \text{ in } L_2^{1,\ell oc}(\Omega)\ .$$

Each is a closed subspace of the Fréchet space $L_2^{1,\ell oc}(\Omega)$. The sets

$$(3.18) \qquad \begin{cases} L_2^{1,p,com}(\Omega) = L_2^{1,p}(\Omega) \cap L_2^{com}(\Omega) \\ \\ L_2^{D,p,com}(\Omega) = L_2^{D,p}(\Omega) \cap L_2^{com}(\Omega) \end{cases}$$

will also be used.

The operator $A_p^{N,\ell oc}(\Omega)$ is the extension of $A_p^N(\Omega)$ in $L_2^{\ell oc}(\Omega)$ defined by

$$(3.19) \qquad D(A_p^{N,\ell oc}(\Omega)) = L_2^{1,p,\ell oc}(\Omega) \cap L_2^{1,\ell oc}(\Delta,\Omega) \cap \{u \mid \int_\Omega \{(\Delta u)\overline{v} + \nabla u \cdot \nabla \overline{v}\}\, dX = 0 \text{ for } v \in L_2^{1,p,com}(\Omega)\}$$

and $A_p^{N,\ell oc}(\Omega)u = -\Delta u$. Similarly, $A_p^{D,\ell oc}(\Omega)$ is the extension of $A_p^D(\Omega)$ in $L_2^{\ell oc}(\Omega)$ defined by

$$(3.20) \qquad D(A_p^{D,\ell oc}(\Omega)) = L_2^{D,p,\ell oc}(\Omega) \cap L_2^{1,\ell oc}(\Delta,\Omega) \cap \{u \mid \int_\Omega \{(\Delta u)\overline{v} + \nabla u \cdot \nabla \overline{v}\}\, dX = 0 \text{ for } v \in L_2^{D,p,com}(\Omega)\}$$

and $A_p^{D,\ell oc}(\Omega)u = -\Delta u$. It is easy to verify that $D(A_p^{N,\ell oc}(\Omega))$ and $D(A_p^{D,\ell oc}(\Omega))$ are closed linear subspaces of the Fréchet space $L_2^{1,\ell oc}(\Delta,\Omega)$ and hence are themselves Fréchet spaces.

The reduced grating propagators for the degenerate grating will be denoted by

$$(3.21) \qquad A_{0,p}^N = A_p^N(B_0)\ , \qquad A_{0,p}^D = A_p^D(B_0)\ , \text{ and}$$

$$(3.22) \qquad A_{0,p}^{N,\ell oc} = A_p^{N,\ell oc}(B_0) \ , \qquad A_{0,p}^{D,\ell oc} = A_p^{D,\ell oc}(B_0) \ .$$

Moreover, the condensed notation of §1 will be used; i.e., A_p will be used to denote either $A_p^N(\Omega)$ or $A_p^D(\Omega)$ in stating results valid for both. Similarly, $A_p^{\ell oc}$ will be used to denote $A_p^{N,\ell oc}(\Omega)$ or $A_p^{D,\ell oc}(\Omega)$. In particular, for the degenerate grating the notation $A_{0,p}$ is used for $A_{0,p}^N$ and $A_{0,p}^D$ and $A_{0,p}^{\ell oc}$ is used for $A_{0,p}^{N,\ell oc}$ and $A_{0,p}^{D,\ell oc}$.

Note that all the p-dependent function spaces defined above are periodic functions of p with period 1. It follows that

$$(3.23) \qquad A_{p+m} = A_p \ , \qquad A_{p+m}^{\ell oc} = A_p^{\ell oc} \ \text{for all } m \in Z \ .$$

Hence it will suffice to study A_p and $A_p^{\ell oc}$ for the reduced momenta $p \in (-1/2,1/2]$.

The resolvent set and spectrum of A_p will be denoted by $\rho(A_p)$ and $\sigma(A_p)$ respectively. Clearly $\sigma(A_p) \subset [0,\infty)$ since A_p is selfadjoint and non-negative. In fact, it will be shown that

$$(3.24) \qquad \sigma(A_p) = [p^2,\infty) \ \text{for all } p \in (-1/2,1/2] \ .$$

This was proved directly by Alber in the cases considered by him [3]. Here it follows from the eigenfunction expansions for A_p given in §5. $\sigma(A_p)$ is a continuous spectrum which may have embedded eigenvalues. It will be shown in §6 that $\sigma_0(A_p)$, the point spectrum of A_p, is discrete; that is, each interval contains finitely many eigenvalues of A_p and the eigenvalues have finite multiplicity. It is of interest for the applications to diffraction gratings to have criteria for $\sigma_0(A_p)$ to be empty. While completely general criteria are not known it will be shown that the hypotheses of Theorem 2.2 imply $\sigma_0(A_p^D) = \phi$ for all $p \in (-1/2,1/2]$.

Eigenfunction expansions for A_p are derived in §5 by perturbation theory starting from $A_{0,p}$. The expansions for $A_{0,p}$, which are elementary, are recorded here as a starting point for the analysis of A_p. Separation of variables applied to $A_{0,p}^N$ leads to the complete family of generalized eigenfunctions

$$(3.25) \qquad \phi_{0,\pm}^N(X,p+m,q) = \phi_0^N(X,p+m,q) = \frac{1}{\pi}\, e^{i(p+m)\cdot x} \cos qy \ , \ m \in Z \ , \ q > 0$$

where $p \in (-1/2,1/2]$ is fixed. Similarly, for $A_{0,p}^D$ one finds the complete family

$$(3.26) \qquad \phi_{0,\pm}^D(X,p+m,q) = \mp i\, \phi_0^D(X,p+m,q) = \frac{\mp i}{\pi}\, e^{i(p+m)\cdot x} \sin qy, \ m \in Z, \ q > 0 \ .$$

To describe the eigenfunction expansions for $A_{0,p}$ the condensed notation $\phi_{0\pm}(X,p+m,q)$ will be used to denote either $\phi^N_{0\pm}$ or $\phi^D_{0\pm}$. Note that

(3.27) $$\phi_{0\pm}(X,p+m,q) = \psi_{0\pm}(X,p+m,q)\big|_{B_0} ,$$

that is, the generalized eigenfunctions for $A_{0,p}$ are obtained from those of A_0 by restricting X to B_0 and the x-momentum parameter to the lines $p' = p + m$ with $m \in Z$ and $p \in (-1/2,1/2]$ fixed. Classical Plancherel theory implies that if $R_0 = (0,\infty)$ then for all $f \in L_2(B_0)$ the limits

(3.28) $$\tilde{f}_{0\pm}(p+m,q) = \underset{M\to\infty}{\ell.i.m.} \int_{B_{0,M}} \overline{\phi_{0\pm}(X,p+m,q)}\, f(X)\, dX$$

exist in $L_2(R_0)$ for $m \in Z$ and $p \in (-1/2,1/2]$ fixed, where $B_{0,M} = B_0 \cap \{X \mid y < M\}$. Note that the $L_2(R_0)$-convergence refers to the variable q. Moreover, Parseval's formula holds in the form

(3.29) $$\|f\|^2_{L_2(B_0)} = \sum_{m\in Z} \|\tilde{f}_{0,\pm}(p+m,\cdot)\|^2_{L_2(R_0)} .$$

Hence, the sequence

(3.30) $$\{\tilde{f}_{0\pm}(p+m,\cdot) \in L_2(R_0) \mid m \in Z\} \in \sum_{m\in Z} \oplus L_2(R_0)$$

and the operator $\Phi_{0\pm,p} : L_2(B_0) \to \sum_{m\in Z} \oplus L_2(R_0)$, defined by

(3.31) $$\Phi_{0\pm,p} f = \{\tilde{f}_{0\pm}(p+m,\cdot) \mid m \in Z\} ,$$

is an isometry. A more careful application of the Plancherel theory shows that $\Phi_{0\pm,p}$ is unitary. Finally, calculation of the spectral family $\{\Pi_{0,p}(\mu) \mid \mu \geq p^2\}$ for $A_{0,p}$ gives

(3.32) $$\Pi_{0,p}(\mu)\, f(X) = \sum_{(p+m)^2\leq\mu} \int_0^{(\mu-(p+m)^2)^{1/2}} \phi_{0\pm}(X,p+m,q)\, \tilde{f}_{0\pm}(p+m,q)\, dq .$$

In particular, making $\mu \to \infty$ gives the eigenfunction expansion

(3.33) $$f(X) = \underset{M\to\infty}{\ell.i.m.} \sum_{|m|\leq M} \int_0^M \phi_{0\pm}(X,p+m,q)\, \tilde{f}_{0\pm}(p+m,q)\, dq ,$$

convergent in $L_2(B_0)$.

The relationship between the R-B waves for A and the reduced propagators A_p will now be discussed. Note first that if ψ is an R-B surface wave for A with x-momentum $p + m$ $(-1/2 < p \leq 1/2,\ m \in Z)$ and $\omega \notin \{|p+\ell| \mid \ell \in Z\}$ then $\psi \in D(A^{\ell oc})$ and for $y > h$

$$(3.34)\quad \psi(x,y) = \sum_{|p+\ell|>\omega} c_\ell \exp\{i(p+\ell)x\} \exp\{-((p+\ell)^2 - \omega^2)^{1/2}\, y\} .$$

It follows that $\phi(x,y) = \psi(x,y)|_\Omega \in D(A_p)$ and $A_p\phi = \omega^2\phi$. Thus ϕ is an eigenfunction of A_p. To formulate the converse, note that every $\phi \in L_2^{\ell oc}(\Omega)$ has a unique p-periodic extension $\psi \in L_2^{\ell oc}(G)$. It is easy to verify that if

$$(3.35)\qquad \Omega^{(m)} = \Omega + (2\pi m, 0)$$

then for each $m \in Z$ the extension ψ is given by

$$(3.36)\qquad \psi(x,y) = \exp\{2\pi i m p\}\, \phi(x - 2\pi m, y) \text{ for all } (x,y) \in \Omega^{(m)} .$$

This defines ψ in $L_2^{\ell oc}(G)$ because G differs from $\cup_{m\in Z} \Omega^{(m)}$ by a Lebesgue null set. The operator $\mathcal{O}^p : L_2^{\ell oc}(\Omega) \to L_2^{\ell oc}(G)$ defined by (3.36) maps $L_2^{\ell oc}(\Omega)$ one-to-one onto the set of all p-periodic functions in $L_2^{\ell oc}(G)$. With this notation it is not difficult to show that if ϕ is an eigenfunction of A_p then $\psi = \mathcal{O}^p\phi$ is an R-B surface wave for A with reduced x-momentum p.

The relationship between R-B diffracted plane waves for A and generalized eigenfunctions of A_p is exemplified by (3.27). More generally, if $\psi(X,p+m,q)$ is an R-B diffracted plane wave for A with $-1/2 < p \leq 1/2$, $m \in Z$ then $\phi_\pm(X,p+m,q) = \psi_\pm(X,p+m,q)|_\Omega$ satisfies $\phi_\pm(\cdot,p+m,q) \in D(A_p^{\ell oc})$, $(\Delta + \omega^2(p+m,q))\, \phi_\pm(X,p+m,q) = 0$ in Ω and

$$(3.37)\qquad \phi_\pm(X,p+m,q) = \phi_{0\pm}(X,p+m,q) + \phi_\pm'(X,p+m,q) , \quad y \geq h ,$$

where ϕ_+' (resp., ϕ_-') has a Fourier expansion that contains only outgoing (resp., incoming) plane waves and exponentially damped waves. Functions $\phi_+(X,p+m,q)$ and $\phi_-(X,p+m,q)$ with these properties will be called, respectively, outgoing and incoming diffracted plane waves for A_p. They are unique modulo eigenfunctions of A_p. It is now easy to verify that if $\phi_+(X,p+m,q)$ (resp., $\phi_-(X,p+m,q)$) is an outgoing (resp., incoming) diffracted plane wave for A_p then $\psi_+(X,p+m,q) = \mathcal{O}^p\, \phi_+(X,p+m,q)$ (resp., $\psi_-(X,p+m,q) = \mathcal{O}^p\, \phi_-(X,p+m,q)$) is an outgoing (resp., incoming) R-B diffracted plane wave for A with x-momentum p + m. These relationships will be used in §8 to construct the R-B diffracted plane waves for A.

§4. Analytic Continuation of the Resolvent of A_p

An analytic continuation of the resolvent

$$(4.1)\qquad R(A_p,z) = (A_p - z)^{-1}$$

across the spectrum $\sigma(A_p) = [p^2,\infty)$ is constructed in this section by an elegant and powerful method that was introduced into scattering theory by H. D. Alber [3]. The continuation provides the basis for the construction in §6 of the diffracted plane waves $\phi_\pm(X,p,q)$ for A_p and the derivation of the corresponding eigenfunction expansions.

The results in this chapter form the core of the analytic theory needed to construct R-B waves and prove their completeness. The proofs of these results offered here are long and arduous, necessarily so in the author's opinion. Therefore, to make the exposition more readable only sketches of proofs are indicated in §4. The complete proofs are given in the following §5.

For each pair of extended real numbers r,r' satisfying $0 \leq r < r' \leq +\infty$ let

$$B_{r,r'} = \{X \mid -\pi < x < \pi,\ r < y < r'\}\ ,\ B_r = B_{r,\infty}\ , \tag{4.2}$$
$$\Omega_{r,r'} = \Omega \cap B_{r,r'}\ ,\ \Omega_r = \Omega_{r,\infty}\ .$$

Moreover, let $P_r : L_2(\Omega_{0,r}) \to L_2(\Omega)$ denote the linear operator defined by

$$P_r\, u(X) = \begin{cases} u(X)\ , & X \in \Omega_{0,r} \\ 0\ , & X \in \Omega_r\ . \end{cases} \tag{4.3}$$

The goal of §4 may be formulated with this notation. It is to construct an analytic continuation of

$$z \to R(A_p,z)\ P_r\ :\ L_2(\Omega_{0,r}) \to L_2^{\ell oc}(\Omega) \tag{4.4}$$

from the resolvent set $\rho(A_p) = C - [p^2,\infty)$ across $\sigma(A_p) = [p^2,\infty)$. For this purpose $\rho(A_p)$ will be embedded in a Riemann surface M_p.

The definition of M_p may be motivated by considering the linear space of functions

$$E_{p,z,r} = D(A_p^{\ell oc}) \cap \{u \mid \text{supp}\ (\Delta + z)u \subset \Omega_{0,r}\}\ ,\ r > h\ . \tag{4.5}$$

Basic properties of $E_{p,z,r}$ are described by

<u>Lemma 4.1</u>. Assume that $\overline{R_h^2} \subset G$. Then every $u \in E_{p,z,r}$ satisfies

$$u \in L_2^{2,\ell oc}(\Omega_h) \tag{4.6}$$

(4.7) $$u(x,y) = \sum_{m \in Z} u_m(y)\, e^{i(p+m)x} \text{ in } \Omega_h$$

where the series converges in $L_2^{2,\ell oc}(\Omega_h)$,

(4.8) $$u_m(y) \in L_2^{2,\ell oc}(R_h) \text{ , } R_h = (h,\infty) \text{ .}$$

Moreover, if $\overline{\Omega}_r$ denotes the closure of Ω_r in R^2,

(4.9) $$u \in C^\infty(\overline{\Omega}_r) \text{ , and}$$

(4.10) $$u_m(y) = c_m^+ \exp\{iy(z-(p+m)^2)^{1/2}\} + c_m^- \exp\{-iy(z-(p+m)^2)^{1/2}\}$$

for $y \geq r$ where $c_m^\pm$ are constants and

(4.11) $$\operatorname{Im}(z-(p+m)^2)^{1/2} \geq 0 \text{ .}$$

Properties (4.6) and (4.9) follow from elliptic regularity theory [1], while (4.7) and (4.8) follow from classical Fourier theory. The convergence of (4.7) in $L_2^{2,\ell oc}(\Omega_h)$ follows from the fact that the partial sums of the Fourier series define orthogonal projections in $L_2^2(\Omega_{r,r'})$ for $h \leq r < r' < \infty$. (4.10) follows from (4.9) and the equation $\Delta u + zu = 0$ in Ω_r.

Note that if $z \in \rho(A_p)$ and $u = R(A_p,z)\, P_r f$ with $f \in L_2(\Omega_{0,r})$ then $u \in L_2(\Omega) \cap E_{p,z,r}$ and hence $c_m^- = 0$ for all m and $c_m^+ = 0$ when $\operatorname{Im}(z-(p+m)^2)^{1/2} = 0$. This suggests that M_p be defined as the Riemann surface associated with the family of holomorphic functions on $C - [p^2,\infty)$ defined by

(4.12) $$\{z \to (z-(p+m)^2)^{1/2} \mid \operatorname{Im}(z-(p+m)^2)^{1/2} > 0 \text{ for all } m \in Z\} \text{ .}$$

M_p is uniquely determined up to isomorphism by the following three properties [3, 16]:

(4.13) M_p is connected and every function of the family (4.12) can be continued analytically to all of M_p.

(4.14) For every pair of points of M_p that lie over the same point of C there are at least two functions of the family that take different values at these points.

(4.15) M_p is maximal with respect to these two properties.

The following notation will be used in connection with M_p. ζ will denote a generic point of M_p and $\pi = \pi_p : M_p \to C$ will denote the canonical projection of M_p onto C. The subscript p will be omitted when there is no danger of ambiguity. The analytic continuation of $(z-(p+m)^2)^{1/2}$ from $C - [p^2,\infty)$ to M_p will be denoted by $w_{p+m}(\zeta)$. Thus, for all $\zeta \in M_p$,

$$w_{p+m}(\zeta) = \pm(\pi(\zeta) - (p+m)^2)^{1/2} \quad . \tag{4.16}$$

M_p^+ will denote that component of M_p over $C - [p^2,\infty)$ on which $\mathrm{Im}\, w_{p+m}(\zeta) > 0$ for all $m \in Z$. Finally, $T_p = \{(p+m)^2 \mid m \in Z\} \subset C$ will denote the set of branch points of the family (4.12).

The properties of M_p include the following. M_p has infinitely many sheets. More precisely, for each disk $D(z_0,\rho) \subset C$, $\pi^{-1}(D(z_0,\rho))$ has infinitely many components. If $z_0 = (p+m)^2$ for some $m \in Z$ then the set $\pi^{-1}(D(z_0,\rho))$ contains infinitely many branch points. Moreover, for all $\zeta \in M_p$ the set $\{m \mid \mathrm{Im}\, w_{p+m}(\zeta) \le 0\}$ is finite [3]. Finally, $M_{p+m} = M_p$ for all $m \in Z$.

In addition to M_p the set

$$M = \bigcup_{-1/2<p\le 1/2} \{(p,\zeta) \mid \zeta \in M_p\} \tag{4.17}$$

will be needed to describe the dependence of the continuation of $R(A_p,\pi(\zeta))\,P_r$ on p and ζ. M will be topologized in such a way that each function $(p,\zeta) \to w_{p+m}(\zeta)$, $m \in Z$, is continuous on M. To this end let $(p_0,\zeta_0) \in M$ and define

$$z_0 = \pi_{p_0}(\zeta_0) \ , \ D(z_0,\rho) = \{z \mid |z - z_0| < \rho\} \tag{4.18}$$

and

$$U(p_0,\zeta_0,\rho) = \text{Component of } \pi_{p_0}^{-1}(D(z_0,\rho)) \text{ containing } \zeta_0 \ (\subset M_{p_0}) \ . \tag{4.19}$$

To define a neighborhood basis for M at (p_0,ζ_0) three cases will be distinguished.

<u>Case 1</u>. $z_0 \notin [p_0^2,\infty)$. If $\rho_0 > 0$ is the distance from z_0 to $[p_0^2,\infty)$ then for $\rho < \rho_0$ $D(z_0,\rho) \cap [p_0^2,\infty) = \phi$ and $U(p_0,\zeta_0,\rho)$ contains no branch points of M_{p_0}. In this case

$$\{\mathrm{sgn}\, \mathrm{Im}\, w_{p_0+m}(\zeta) \mid m \in Z\} \ , \ \zeta \in U(p_0,\zeta_0,\rho) \tag{4.20}$$

is well defined. Moreover, $|p - p_0| < \delta$ implies that $D(z_0,\rho) \cap [p^2,\infty) = \phi$ for δ small enough and hence $\{\mathrm{sgn}\, \mathrm{Im}\, w_{p+m}(\zeta) \mid m \in Z\}$ is also well defined on the components of $\pi_p^{-1}(D(z_0,\rho))$. In this case one may define $U(p,\zeta_0,\rho)$ as the component of $\pi_p^{-1}(D(z_0,\rho))$ for which

(4.21) $$\{\operatorname{sgn} \operatorname{Im} w_{p+m}(\pi_p^{-1}(z)) \mid m \in Z\} = \{\operatorname{sgn} \operatorname{Im} w_{p_0+m}(\zeta_0) \mid m \in Z\}$$

for $z \in D(z_0,\rho)$. A corresponding neighborhood of (p_0,ζ_0) in M is defined by

(4.22) $$N(p_0,\zeta_0,\rho,\delta) = \bigcup_{|p-p_0|<\delta} \{(p,\zeta) \mid \zeta \in U(p,\zeta_0,\rho)\} .$$

<u>Case 2</u>. $z_0 \in [p^2,\infty) - T_{p_0}$. In this case if ρ_0 is the distance from z_0 to the set T_{p_0} then for $\rho < \rho_0$ $U(p_0,\zeta_0,\rho)$ contains no branch points of M_{p_0} and (4.20) is well defined provided $\pi_{p_0}(\zeta) \in D_+(z_0,\rho) = D(z_0,\rho) \cap \{z \mid \operatorname{Im} z > 0\}$. Moreover, $|p - p_0| < \delta$ implies that $D(z_0,\rho)$ contains no points of T_p, for δ small enough, and hence $\{\operatorname{sgn} \operatorname{Im} w_{p+m}(\zeta) \mid m \in Z\}$ is also well defined if $\pi_p(\zeta) \in D_+(z_0,\rho)$. In this case one defines $U(p,\zeta_0,\rho)$ as the component of $\pi_p^{-1}(D(z_0,\rho))$ for which (4.21) holds for $z \in D_+(z_0,\rho))$. A corresponding neighborhood is again defined by (4.22).

<u>Case 3</u>. $z_0 = (p_0 + m_0)^2$ for some $m_0 \in Z$. If $\rho_0 > 0$ is the distance from z_0 to the set $T_{p_0} - \{(p_0 + m_0)^2\}$ then for $\rho < \rho_0$ the set $U(p_0,\zeta_0,\rho)$ contains only one branch point; namely, that for $w_{p_0+m_0}(\zeta)$. Hence $\{\operatorname{sgn} \operatorname{Im} w_{p_0+m}(\zeta) \mid m \in Z - \{m_0\}\}$ is well defined for $\zeta \in U(p_0,\zeta_0,\rho)$ and $\pi_{p_0}(\zeta) \in D_+(z_0,\rho)$. Moreover, $|p - p_0| < \delta$ implies that $D(z_0,\rho)$ contains $(p + m_0)^2$ and no other points of the set T_p and hence $\{\operatorname{sgn} \operatorname{Im} w_{p+m}(\zeta) \mid m \in Z - \{m_0\}\}$ is well defined on the components of $\pi_p^{-1}(D_+(z_0,\rho))$. In this case one may define $U(p,\zeta_0,\rho)$ as the component of $\pi_p^{-1}(D(z_0,\rho))$ for which

(4.23) $$\{\operatorname{sgn} \operatorname{Im} w_{p+m}(\pi_p^{-1}(z)) \mid m \in Z - \{m_0\}\} = \{\operatorname{sgn} \operatorname{Im} w_{p_0+m}(\pi_{p_0}^{-1}(z)) \mid m \in Z - \{m_0\}\}$$

for all $z \in D_+(z_0,\rho)$. A corresponding neighborhood is again defined by (4.22).

The topology of M is defined to be the one generated by the neighborhood bases defined above and one has

<u>Theorem 4.2</u>. Each of the functions on M defined by

(4.24) $$(p,\zeta) \to w_{p+m}(\zeta) \ , \ m \in Z$$

is continuous on M. Moreover, the family of functions $\{(p,\zeta) \to w_{p+m}(\zeta) \mid m \in Z\}$ is equicontinuous in M.

The theorem that $\{\zeta \to w_{p+m}(\zeta) \mid m \in Z\}$ is equicontinuous on M_p for fixed p was proved by Alber [3]. Theorem 4.2 plays a key role in proving the continuity in (p,q) of the Rayleigh-Bloch waves in §8.

<u>The Fréchet Space $F_{p,\zeta,r}$</u>. To describe the subset of $E_{p,z,r}$ that contains the analytic continuation of $R(A_p,z)\, P_r f$ to M_p, consider the set of functions $u \in E_{p,z,r}$ whose Fourier representations (4.7), (4.10) in Ω_r satisfy

$$\text{(4.25)} \qquad \text{For each } m \in Z \text{, either } c_m^+ = 0 \text{ or } c_m^- = 0 \text{, and}$$

$$\text{(4.26)} \qquad c_m^- = 0 \text{ for all but a finite number of } m \in Z .$$

Note that these conditions express the "radiation conditions"

$$\text{(4.27)} \qquad [D_y \pm i(z - (p+m)^2)^{1/2}]\, u_m(y) = 0 \text{ , } y \geq r \text{ ,}$$

where for each m either "+" or "-" is chosen and "-" is chosen for all but a finite number of $m \in Z$. It is clear that each such $u \in E_{p,z,r}$ is associated with a unique point $\zeta \in M_p$ such that $\pi_p(\zeta) = z$ and the Fourier expansion (4.7), (4.10) of u has the form

$$\text{(4.28)} \qquad u(x,y) = \sum_{m \in Z} c_m \exp\{i(p+m)x + i y\, w_{p+m}(\zeta)\} \text{ , } y \geq r \text{ .}$$

For each $(p,\zeta) \in M$ and each $r > h$ the set of all such solutions will be denoted by

$$\text{(4.29)} \quad F_{p,\zeta,r} = D(A_p^{\ell oc}) \cap \{u \mid \operatorname{supp}\, (\Delta + \pi_p(\zeta))u \subset \Omega_{0,r} \text{ and (4.28) holds}\} \text{ .}$$

Note that $F_{p,\zeta,r} \subset D(A_p^{\ell oc}) \subset L_2^{1,\ell oc}(\Delta,\Omega)$ and recall that $D(A_p^{\ell oc})$ is closed in $L_2^{1,\ell oc}(\Delta,\Omega)$. This implies

<u>Theorem 4.3</u>. $F_{p,\zeta,r}$ is closed in $D(A_p^{\ell oc})$ in the topology of $L_2^{1,\ell oc}(\Delta,\Omega)$ and hence is a Fréchet space.

This is immediate because the defining properties of $F_{p,\zeta,r}$, namely $\operatorname{supp}\, (\Delta + \pi_p(\zeta))u \subset \Omega_{0,r}$ and $[D_y - i\, w_{p+m}(\zeta)]\, u_m = 0$ in $y \geq r$, are preserved under convergence in $L_2^{1,\ell oc}(\Delta,\Omega)$.

The following condensed notation will be used in discussing $F_{p,\zeta,r}$ and related operators:

$$\text{(4.30)} \qquad (u,v)_{r,r'} = (u,v)_{L_2(\Omega_{r,r'})} \text{ ,}$$

(4.31) $$(u,v)_{1;r,r'} = (u,v)_{L_2^1(\Omega_{r,r'})} ,$$

(4.32) $$(u,v)_{1;\Delta;r,r'} = (u,v)_{L_2^1(\Delta,\Omega_{r,r'})} .$$

Now let $P_{p,\zeta,r} : F_{p,\zeta,r} \to L_2(\Omega_{0,r})$ denote the natural projection defined by

(4.33) $$P_{p,\zeta,r}\, u = u|_{\Omega_{0,r}} \quad \text{for all } u \in F_{p,\zeta,r} .$$

An important property of $F_{p,\zeta,r}$ is expressed by the following generalization of a theorem of Alber [3, p. 264].

Theorem 4.4. For every compact set $K \subset M$ and for every $r' > r$ there exists a constant $C = C(K,r,r')$ such that

(4.34) $$\|u\|_{1;\Delta;0,r'} \le C\, \|P_{p,\zeta,r}\, u\|_{1;\Delta;0,r}$$

for all $u \in \bigcup_{(p,\zeta)\in K} F_{p,\zeta,r}$. In particular, $P_{p,\zeta,r}$ is a topological isomorphism of $F_{p,\zeta,r}$ onto $P_{p,\zeta,r}\, F_{p,\zeta,r}$, topologized by the $L_2^1(\Delta,\Omega_{0,r})$-norm.

The Operators $A_{p,\zeta,r} : L_2(\Omega_{0,r}) \to L_2(\Omega_{0,r})$. Following Alber's program, the construction of the analytic continuation of $R(A_p,z)\, P_r$ to M_p will be based on the family of linear operators $A_{p,\zeta,r}$ in $L_2(\Omega_{0,r})$, defined for all $(p,\zeta) \in M$ by

(4.35) $$D(A_{p,\zeta,r}) = P_{p,\zeta,r}\, F_{p,\zeta,r} ,$$

(4.36) $$A_{p,\zeta,r}\, u = -\Delta u .$$

The properties of $A_{p,\zeta,r}$ that are fundamental for the analytic continuation of $R(A_p,z)$ are described by the following theorems.

Theorem 4.5. For every $(p,\zeta) \in M$ and every $r > h$ the operator $A_{p,\zeta,r}$ is m-sectorial in the sense of Kato [13, p. 279].

Theorem 4.6. For all grating domains of the class defined in §1, the family of operators $\{A_{p,\zeta,r} \mid (p,\zeta) \in M\}$ is continuous in the sense of generalized convergence (Kato [13, p. 206]). Moreover, for each fixed $p \in (-1/2,1/2]$ the family $\{A_{p,\zeta,r} \mid \zeta \in M_p\}$ is holomorphic in the generalized sense (Kato [13, p. 366]).

Theorem 4.7. For every $(p,\zeta) \in M$, every $r > h$ and every $z \in \rho(A_{p,\zeta,r})$ the resolvent $R(A_{p,\zeta,r},z) = (A_{p,\zeta,r} - z)^{-1}$ is a compact operator in $L_2(\Omega_{0,r})$ and hence $\sigma(A_{p,\zeta,r})$ is discrete.

Theorem 4.5 generalizes Alber [3, Th. 5.5]. As in [3] it may be proved by associating $A_{p,\zeta,r}$ with a densely defined, closed, sectorial sesquilinear form in $L_2(\Omega_{0,r})$ and using Kato's first representation theorem [13, p. 322]. The second statement of Theorem 4.6 generalizes Alber [3, Th. 5.5b]. The hypothesis $G \in S$ of §1 is needed to prove Theorem 4.6. Theorem 4.7, which generalizes Alber [3, Th. 5.5a], is a consequence of the local compactness property of G in the case of the Neumann boundary conditions. Complete proofs of Theorems 4.5, 4.6 and 4.7 are given in §5. The following consequences of these theorems are needed for the spectral analyses of A_p and A in §6 and §8.

Theorem 4.8. For all $\zeta \in M_p^+$ one has $\pi_p(\zeta) \in \rho(A_{p,\zeta,r})$ and

$$R(A_{p,\zeta,r},\pi_p(\zeta)) = P_{p,\zeta,r}\, R(A_p,\pi_p(\zeta))\, P_r \; . \tag{4.37}$$

This result may be verified by direct calculation.

Theorem 4.9. For every $p \in (-1/2,1/2]$ the set

$$\Sigma_p = \{\zeta \mid \pi_p(\zeta) \in \sigma(A_{p,\zeta,r})\} \subset M_p \tag{4.38}$$

has no accumulation points in M_p and is independent of $r > h$.

This result, which generalizes [3, Th. 5.5c], is a consequence of Theorem 4.7. For brevity the resolvent of (4.37) will be denoted by

$$R_{p,\zeta,r} = R(A_{p,\zeta,r},\pi_p(\zeta)) \in B(L_2(\Omega_{0,r})) \; . \tag{4.39}$$

Here B(X) denotes the bounded operators on X.

Corollary 4.10. For each $p \in (-1/2,1/2]$ and $r > h$ the mapping

$$\zeta \to R_{p,\zeta,r} \in B(L_2(\Omega_{0,r})) \tag{4.40}$$

is finitely meromorphic on M_p with pole set Σ_p.

This result is based on a theorem of S. Steinberg [24]; cf. [3, Th. 5.5e]. Theorem 4.4 and 4.8 provide the analytic continuation of $R(A_p,z)\, P_r$ in the following form.

Corollary 4.11. The analytic continuation to M_p of

$$\zeta \to R(A_p,\pi_p(\zeta))\, P_r \in B(L_2(\Omega_{0,r}) \; , \; L_2^{1,\ell oc}(\Delta,\Omega)) \; , \; \zeta \in M_p^+ \; , \tag{4.41}$$

is given by

$$(4.42) \qquad \zeta \to P^{-1}_{p,\zeta,r} R_{p,\zeta,r} \in B(L_2(\Omega_{0,r}), L_2^{1,\ell oc}(\Delta,\Omega)), \ \zeta \in M_p ,$$

where B(X,Y) denotes the bounded linear operators from X to Y.

Corollary 4.12. For all grating domains of the class defined in §1, the point spectrum $\sigma_0(A_p)$ is discrete.

This result follows from Theorem 4.9 and Corollary 4.10.

Corollary 4.13. For all grating domains of the class defined in §1 one has

$$(4.43) \qquad \pi_p\left(\overline{M_p^+} \cap \Sigma_p\right) \subset \sigma_0(A_p) \cup T_p$$

where $\overline{M_p^+}$ is the closure of M_p^+ in M_p.

If $\sigma_0(A_p) = \phi$ then (4.43) means that the poles of $R_{p,\zeta,r}$ that lie above the spectrum $\sigma(A_p) = [p^2,\infty)$ must lie above the branch point set T_p. The fact that such poles may or may not occur is illustrated by the two operators A_0^D and A_0^N corresponding to the degenerate grating. For A_0^D, separation of variables leads to a construction of the Green's function (= kernel of the resolvent $R(A_0^D,a)$) which can be written

$$G_0^D(X,X',p,z)$$

$$(4.44)$$

$$= \frac{i}{2\pi} \sum_{m\in Z} e^{i(p+m)(x-x')}(z-(p+m)^2)^{-1/2} \sin\, (z-(p+m)^2)^{1/2}\, y_< \, e^{i(z-(p+m)^2)^{1/2}\, y_>}$$

where $y_< = \text{Min}\,(y,y')$, $y_> = \text{Max}\,(y,y')$. The analogous calculation for A_0^N gives

$$G_0^N(X,X',p,z)$$

$$(4.45)$$

$$= \frac{i}{2\pi} \sum_{m\in Z} e^{i(p+m)(x-x')}(z-(p+m)^2)^{-1/2} \cos\, (z-(p+m)^2)^{1/2}\, y_< \, e^{i(z-(p+m)^2)^{1/2}\, y_>} .$$

In the first case $R(A_0^D,z)$ has no poles for real $z = \lambda \pm i0 \in [p^2,\infty)$. In the second case $R(A_0^N,z)$ has a simple pole at each of the points $z = \lambda \pm i0 \in T_p$.

The following two theorems are implied by Theorems 4.4 and 4.6.

Theorem 4.14. Let

$$\Sigma = \bigcup_{-1/2<p\leq 1/2} \{(p,\zeta) \mid \zeta \in \Sigma_p\} = \bigcup_{-1/2<p\leq 1/2} \{(p,\zeta) \mid \pi_p(\zeta) \in \sigma(A_{p,\zeta,r})\} .$$

(4.46)

Then $M-\Sigma$ is open in M and

$$(p,\zeta) \rightarrow R_{p,\zeta,r} \in B(L_2(\Omega_{0,r})) \tag{4.47}$$

is continuous on $M-\Sigma$.

Theorem 4.15. The mapping

$$(p,\zeta) \rightarrow P^{-1}_{p,\zeta,r}\, R_{p,\zeta,r} \in B(L_2(\Omega_{0,r})\ ,\ L_2^{1,\ell oc}(\Delta,\Omega)) \tag{4.48}$$

is continuous on $M-\Sigma$.

A direct consequence of Theorem 4.15 that is needed below is

Corollary 4.16. Let K be any compact subset of $M-\Sigma$ and let $r' > r > h$. Then there exists a constant $C = C(K,r,r')$ such that

$$\|P^{-1}_{p,\zeta,r}\, R_{p,\zeta,r}\, f\|_{1;\Delta;0,r'} \leq C\, \|f\|_{0,r} \tag{4.49}$$

for all $(p,\zeta) \in K$ and all $f \in L_2(\Omega_{0,r})$.

A Limiting Absorption Theorem. In the remainder of this work the point ζ will be restricted to $\overline{M_p^+}$, the closure of M_p^+ in M_p. To simplify the notation points $\zeta \in M_p^+$ will be identified with their images $\pi_p(\zeta)$ $= z \in C - [p^2,\infty)$ and the points of ∂M_p^+ will be denoted by $\lambda \pm i0$, where $\lambda \in [p^2,\infty)$. With this notation the operators

$$P^{-1}_{p,\lambda\pm i\sigma,r}\, R_{p,\lambda\pm i\sigma,r} \in B(L_2(\Omega_{0,r}),L_2^{1,\ell oc}(\Delta,\Omega)) \tag{4.50}$$

are defined and continuous for all $\lambda \pm i\sigma \in \overline{M_p^+} - \Sigma_p$. Note that by Corollary 4.13

$$\sigma(A_p) - \sigma_0(A_p) - T_p \subset \pi_p(\partial M_p^+ - \Sigma_p) . \tag{4.51}$$

Now let $f \in L_2(\Omega_{0,r})$ and define

$$u_\pm(\cdot,p,\lambda) = P^{-1}_{p,\lambda\pm i0,r}\, R_{p,\lambda\pm i0,r}\, f \in F_{p,\lambda\pm i0,r} . \tag{4.52}$$

Then, in particular,

$$u_\pm(\cdot,p,\lambda) \in D(A_p^{\ell oc})\ , \text{ and} \tag{4.53}$$

(4.54) $$\Delta u_\pm + \lambda u_\pm = f \text{ in } \Omega.$$

Moreover, $\pi_p(\lambda \pm i0) = \lambda$ for all $\lambda \in \sigma(A_p)$ and

(4.55)
$$w_{p+m}(\lambda \pm i0) = \pm(\lambda - (p+m)^2)^{1/2} \quad \text{if } \lambda > (p+m)^2 ,$$
$$w_{p+m}(\lambda \pm i0) = i((p+m)^2 - \lambda)^{1/2} \quad \text{if } \lambda < (p+m)^2 .$$

Hence, the Fourier series (4.28) of $u_\pm$ have the form

(4.56)
$$u_\pm(x,y,p,\lambda) = \sum_{(p+m)^2<\lambda} c_m^\pm e^{i(p+m)x} e^{\pm iy(\lambda-(p+m)^2)^{1/2}}$$
$$+ \sum_{(p+m)^2>\lambda} c_m^\pm e^{i(p+m)x} e^{-y((p+m)^2-\lambda)^{1/2}} .$$

Thus u_+ and u_- are the outgoing and incoming solutions, respectively, of the boundary value problem (4.53), (4.54). Moreover, they are uniquely determined by these conditions, by Theorem 2.1, provided

(4.57) $$\lambda \in \sigma(A_p) - \sigma_0(A_p) - T_p .$$

The final result of this section is a uniform bound for the functions

(4.58) $$P^{-1}_{p,\lambda\pm i\sigma,r} R_{p,\lambda\pm i\sigma,r} f \in L_2^{1,\ell oc}(\Delta,\Omega)$$

which may be formulated as follows.

Corollary 4.17. Let $I = [a,b]$ satisfy

(4.59) $$I \subset \sigma(A_p) - \sigma_0(A_p) - T_p$$

and let p, σ_0, r and r' satisfy $-1/2 < p \le 1/2$, $\sigma_0 > 0$ and $r' > r > h$. Then there exists a constant $C = C(I,p,\sigma_0,r,r')$ such that

(4.60) $$\|P^{-1}_{p,\lambda\pm i\sigma,r} R_{p,\lambda\pm i\sigma,r} f\|_{1;\Delta;0,r'} \le C \|f\|_{0,r}$$

for all $\lambda \in I$, $0 \le \sigma \le \sigma_0$ and all $f \in L_2(\Omega_{0,r})$. Moreover, if $\Sigma_p \cap \sigma(A_p) = \phi$ then the same result holds for intervals $I \subset \sigma(A_p) - \sigma_0(A_p)$.

§5. Proofs of the Results of §4.

Proof of Lemma 4.1. Assume that $u \in D(A_p^{\ell oc})$ and define $v(x,y) = \exp\{-ipx\} u(x,y)$. Then $v \in L_2^{1,\ell oc}(\Delta,\Omega)$ and satisfies the p-periodic boundary conditions (3.7) with $p = 0$. Thus if Ω^γ is the cylinder obtained by identifying the points $(-\pi,y)$ and (π,y), $y \in \gamma$, it follows that v is a distribution solution of $\Delta v + 2ip\, D_x v + (z-p^2)v = e^{-ipx} f \in L_2^{\ell oc}(\Omega^\gamma)$. Let h' satisfy $0 < h' < h$, $R^2_{h'} \subset G$. Such numbers h' exist if $\overline{R^2_h} \subset G$. Then $\Omega^\gamma_{h'} = \Omega^\gamma \cap \{(x,y) \mid y > h'\}$ is contained in the interior of Ω^γ and the interior elliptic estimates of [1] imply that $v \in L_2^{2,\ell oc}(\Omega^\gamma_{h'})$. This result implies (4.6) and (4.8) of Lemma 4.1. Moreover, $f = 0$ in Ω_r and the regularity theory of [1] implies $v \in L_2^{m,\ell oc}(\Omega^\gamma_r)$ for all $m \in Z$ which implies (4.9) and (4.10).

It remains to prove (4.7). Note that if v is defined as above then

(5.1) $$u_m(y) = \frac{1}{2\pi} \int_{-\pi}^{\pi} \overline{e^{i(p+m)x}}\, u(x,y)\,dx = \frac{1}{2\pi} \int_{-\pi}^{\pi} \overline{e^{imx}}\, v(x,y)\,dx = v_m(y) .$$

Hence (4.7) is equivalent to the statement that

(5.2) $$v(x,y) = \sum_{m\in Z} v_m(y)\, e^{imx} \text{ in } \Omega^\gamma_h$$

where the series converges to v in $L_2^{2,\ell oc}(\Omega^\gamma_h)$. To prove this note that $\{e^{imx} \mid m \in Z\}$ is an orthogonal sequence in $L_2^2(\Omega^\gamma_{k,k'})$ for any k, k' such that $h \leq k < k' < \infty$. Next define

(5.3) $$P_\ell\, v(x,y) = \sum_{|m|\leq\ell} v_m(y)\, e^{imx}$$

where v_m is defined by (5.1). Then direct calculation shows that

(5.4) $$P_\ell : L_2^2(\Omega^\gamma_{k,k'}) \to L_2^2(\Omega^\gamma_{k,k'}) \text{ is bounded}$$

and

(5.5) $$P_\ell^2 = P_\ell = P_\ell^* \text{ in } L_2^2(\Omega^\gamma_{k,k'}) ;$$

i.e., P_ℓ is an orthogonal projection. It follows that

(5.6) $$Q_\ell = 1 - P_\ell$$

is also an orthogonal projection in $L_2^2(\Omega^\gamma_{k,k'})$. Note that the convergence of (5.2) in $L_2^{2,\ell oc}(\Omega^\gamma_{k,k'})$ is equivalent to the condition

(5.7) $$\lim_{\ell\to\infty} \|Q_\ell\, v\|_2 = 0 \text{ for all } v \in L_2^2(\Omega^\gamma_{k,k'})$$

where $\|\cdot\|_2$ is the norm in $L_2^2(\Omega^\gamma_{k,k'})$. Now (5.7) follows from classical convergence theory for Fourier series if $v \in C^\infty(\Omega^\gamma_{k,k'})$, the set of restrictions to $\overline{\Omega^{\gamma}_{k,k'}}$ of functions from $C^{\infty}(\Omega^{\gamma}_{h})$. Moreover, this set is dense in $L^2_2(\Omega^{\gamma}_{k,k'})$. Thus if $v \in L^2_2(\Omega^{\gamma}_{k,k'})$ and $v' \in C^{\infty}(\overline{\Omega^{\gamma}_{k,k'}})$ then

$$\|Q_{\ell}\, v\|_2^2 = (Q_{\ell}\, v,v)_2 = (Q_{\ell}(v-v'),v)_2 + (Q_{\ell}\, v',v)_2$$

(5.8)

$$\leq \|v-v'\|_2\, \|v\|_2 + \|Q_{\ell}\, v'\|_2\, \|v\|_2$$

It follows that

(5.9) $$\limsup_{\ell\to\infty} \|Q_{\ell}\, v\|_2^2 \leq \|v-v'\|_2\, \|v\|_2$$

for all $v' \in C^{\infty}(\overline{\Omega^{\gamma}_{k,k'}})$ which implies (5.7).

Proof of Theorem 4.2. To prove the continuity of the mappings $(p,\zeta) \to w_{p+m}(\zeta)$ for all $(p,\zeta) \in M$ and $m \in Z$ let $(p_0,\zeta_0) \in M$, $m \in Z$ and $\varepsilon > 0$. It will be shown that there exist $\rho_0(\varepsilon) > 0$ and $\delta_0(\varepsilon) > 0$ such that

(5.10) $$|w_{p+m}(\zeta) - w_{p_0+m}(\zeta_0)| < \varepsilon \text{ for } (p,\zeta) \in N(p_0,\zeta_0,\rho,\delta)$$

provided $0 < \rho \leq \rho_0(\varepsilon)$, $0 < \delta \leq \delta_0(\varepsilon)$.

To prove (5.10) note that in Cases 1 and 2 of the definition of $N(p_0,\zeta_0,\rho,\delta)$ one has, for every $m \in Z$,

(5.11) $$w_{p+m}(\zeta) = \pm(\pi_p(\zeta) - (p+m)^2)^{1/2}\,,\quad w_{p_0+m}(\zeta_0) = \pm(\pi_{p_0}(\zeta_0) - (p_0+m)^2)^{1/2}$$

where the square roots have non-negative imaginary part and the $\pm$ signs are the same for each $m \in Z$. Moreover,

(5.12) $$\pi_p(\zeta),\ \pi_{p_0}(\zeta_0) = z_0 \in D(z_0,\rho) \text{ and } |p - p_0| < \delta$$

for $(p,\zeta) \in N(p_0,\zeta_0,\rho,\delta)$. Hence there exist $\rho_0(\varepsilon) > 0$, $\delta_0(\varepsilon) > 0$ such that

$$|w_{p+m}(\zeta) - w_{p_0+m}(\zeta_0)| = |(\pi_p(\zeta) - (p+m)^2)^{1/2} - (\pi_{p_0}(\zeta_0) - (p_0+m)^2)^{1/2}| < \varepsilon$$

(5.13)

for $(p,\zeta) \in N(p_0,\zeta_0,\rho_0(\varepsilon),\delta_0(\varepsilon))$. To prove (5.10) in Case 3 note that in this case if $(p,\zeta) \in N(p_0,\zeta_0,\rho,\delta)$ then one has both $z_0 = (p_0 + m_0)^2$ and $(p + m_0)^2$ in $D(z_0,\rho)$ for $\delta \leq \delta_0(\rho)$. Moreover $w_{p_0+m_0}(\zeta_0) = 0$. Hence there exists a $\rho_0(\varepsilon) > 0$ such that

(5.14) $$|w_{p+m_0}(\zeta) - w_{p_0+m}(\zeta_0)| = |(\pi_p(\zeta) - (p + m_0)^2)^{1/2}| < \varepsilon$$

for $(p,\zeta) \in N(p_0,\zeta_0,\rho,\delta)$, $0 < \rho \leq \rho_0(\varepsilon)$, $0 < \delta \leq \delta_0(\rho_0(\varepsilon))$ because

$\pi_p(\zeta) \in D(z_0,\rho)$ for all such (p,ζ). The proof that the functions $w_{p+m}(\zeta)$ with $m \neq m_0$ are continuous at (p_0,ζ_0) is the same as in Case 1.

To prove the equicontinuity statement of Theorem 4.2 fix $(p_0,\zeta_0) \in M$. Then for all $m \in Z$ (resp., $m \in Z - \{m_0\}$) Cases 1 or 2 apply to $w_{p+m}(\zeta)$ and if

$$F_m(z,p) = (z-(p+m)^2)^{1/2}\ , \ \operatorname{Im} F_m(z,p) \geq 0 \tag{5.15}$$

then for all $(p,\zeta) \in N(p_0,\zeta_0,\rho,\delta)$ one has

$$|w_{p+m}(\zeta) - w_{p_0+m}(\zeta_0)| = |F_m(\pi_p(\zeta),p) - F_m(\pi_{p_0}(\zeta_0),p_0)|\ . \tag{5.16}$$

Note that $F_m(z,p)$ has partial derivatives

$$\begin{aligned} D_z\, F_m(z,p) &= \frac{1}{2}\,(z-(p+m)^2)^{-1/2}\ , \\ D_p\, F_m(z,p) &= -(z-(p+m)^2)^{-1/2}\,(p+m)\ . \end{aligned} \tag{5.17}$$

Hence for $z \in D(z_0,\rho)$ and $|p - p_0| < \delta$ these derivatives are uniformly bounded for all $m \in Z$ (resp., $m \in Z - \{m_0\}$). Now by Taylor's theorem

$$F_m(z,p) = F_m(z_0,p_0) + (z - z_0)\, D_z\, F_m(z',p') + (p - p_0)\, D_p\, F_m(z',p') \tag{5.18}$$

where (z',p') is on the segment from (z_0,p_0) to (z,p). Thus one has

$$\begin{aligned} &|w_{p+m}(\zeta) - w_{p_0+m}(\zeta_0)| \\ &\leq |D_z\, F_m(z',p')|\ |\pi_p(\zeta) - \pi_{p_0}(\zeta_0)| + |D_p\, F_m(z',p')|\ |p-p_0| \\ &\leq \text{Const.}\ (|\pi_p(\zeta) - \pi_{p_0}(\zeta_0)| + |p - p_0|) \end{aligned} \tag{5.19}$$

for all $(p,\zeta) \in N(p_0,\zeta_0,\rho,\delta)$ and all $m \in Z$ (resp., $m \in Z - \{m_0\}$). Since $\pi_p(\zeta)$ and $\pi_{p_0}(\zeta_0) = z_0$ are in $D(z_0,\rho)$ for $(p,\zeta) \in N(p_0,\zeta_0,\rho,\delta)$ it is clear that there exist $\rho_0(\varepsilon) > 0$, $\delta_0(\varepsilon) > 0$ such that (5.14) holds for all (p,ζ) in $N(p_0,\zeta_0,\rho_0(\varepsilon),\delta_0(\varepsilon))$ and all $m \in Z$.

Theorem 4.3 was proved in §4. The proof of Theorem 4.4 will be based on the following

Lemma 5.1. For every compact set $K \subset M$ and every $r' > r$ there exists a constant $C_1 = C_1(K,r,r')$ such that for all $u \in \cup_{(p,\zeta)\in K} F_{p,\zeta,r}$ one has

$$\|u\|^2_{r,r'} \leq C_1(\|u\|^2_{h,r} + \|\nabla u\|^2_{h,r})\ . \tag{5.20}$$

Proof of Lemma 5.1. Note that every $u \in F_{p,\zeta,r}$ can be written

$$u(X) = u'(X) + u''(X) \ , \ X \in \Omega_r \ , \tag{5.21}$$

where

$$u'(X) = \Sigma' \ c_m \exp \{i\, x(p+m) + i\, y\, w_{p+m}(\zeta)\} \ , \tag{5.22}$$

$$u''(X) = \Sigma'' \ c_m \exp \{i\, x(p+m) + i\, y\, w_{p+m}(\zeta)\} \ , \tag{5.23}$$

the c_m are the coefficients of (4.28) and the notation Σ', Σ'' denotes summation over the index sets $\{m \mid \operatorname{Im} w_{p+m}(\zeta) > 0\}$ and $\{m \mid \operatorname{Im} w_{p+m}(\zeta) \le 0\}$, respectively. Lemma 4.1 implies that $u \in C^\infty(\overline{\Omega}_r)$ and the Fourier series in (5.22) and its derivatives converge uniformly on compact subsets of $\overline{\Omega}_r$ to u' and its derivatives. Moreover, the sum in (5.23) is finite for each $(p,\zeta) \in M$. Finally

$$\|u\|^2_{r,r'} = \|u'\|^2_{r,r'} + \|u''\|^2_{r,r'} \tag{5.24}$$

because $\{e^{i(p+m)x}\}$ is an orthogonal sequence in $L_2(\Omega_{r,r'})$ and the index sets defining Σ' and Σ'' are complementary.

Parseval's relation for Fourier series implies that

$$\int_{-\pi}^{\pi} |u'(x,y)|^2 \, dx = 2\pi \ \Sigma' \ |c_m|^2 \exp \{-2y \operatorname{Im} w_{p+m}(\zeta)\} \tag{5.25}$$

for all $y \ge r$. Moreover, this is a monotone decreasing function of y, whence

$$\int_{-\pi}^{\pi} |u'(x,y)|^2 \, dx \le 2\pi \ \Sigma' \ |c_m|^2 \exp \{-2r \operatorname{Im} w_{p+m}(\zeta)\} = 2\pi\, \Sigma' \ |u_m(r)|^2 \ . \tag{5.26}$$

Integrating this inequality over $r \le y \le r'$ gives

$$\|u'\|_{r,r'} \le 2\pi(r' - r) \ \Sigma' \ |u_m(r)|^2 \ . \tag{5.27}$$

The analogue of (5.25) for u'' is a monotone increasing function of $y \ge r$. In particular, for $r \le y \le r'$ one has

$$\int_{-\pi}^{\pi} |u''(x,y)|^2 \, dx \le 2\pi \ \Sigma'' \ |c_m|^2 \exp \{-2r' \operatorname{Im} w_{p+m}(\zeta)\} \ . \tag{5.28}$$

To estimate this sum note that the sets $\{m \mid \operatorname{Im} w_{p+m}(\zeta) \le 0\}$ vary with $(p,\zeta) \in M$ and the properties of M established in §4 imply that the set

$$M = M(K) = \bigcup_{(p,\zeta)\in K} \{m \mid \operatorname{Im} w_{p+m}(\zeta) \le 0\} \tag{5.29}$$

is finite for each compact $K \subset M$. It follows from this and Theorem 4.2 that

$$\mu = \mu(K) = \text{Max}\ \{-\text{Im}\ w_{p+m}(\zeta) : (p,\zeta) \in K \text{ and } m \in M(K)\} \tag{5.30}$$

is finite. Hence (5.28) implies

$$\begin{aligned} \int_{-\pi}^{\pi} |u''(x,y)|^2\, dx &\le 2\pi \exp\{2(r'-r)\mu\}\ \Sigma''\ |c_m|^2 \exp\{-2r\ \text{Im}\ w_{p+m}(\zeta)\} \\ &= 2\pi \exp\{2(r'-r)\mu\}\ \Sigma''\ |u_m(r)|^2 \end{aligned} \tag{5.31}$$

for $r \le y \le r'$. Integrating (5.31) over $r \le y \le r'$ gives

$$\|u''\|^2_{r,r'} \le 2\pi\ (r'-r) \exp\{2(r'-r)\mu\}\ \Sigma''\ |u_m(r)|^2 . \tag{5.32}$$

Adding (5.27) and (5.32) and using (5.24) gives

$$\|u\|^2_{r,r'} \le 2\pi(r'-r) \exp\{2(r'-r)\mu\} \sum_{m\in Z} |u_m(r)|^2 . \tag{5.33}$$

Finally, Parseval's relation in $L_2(-\pi,\pi)$ gives

$$\|u(\cdot,r)\|^2 = \int_{-\pi}^{\pi} |u(x,r)|^2\, dx = 2\pi \sum_{m\in Z} |u_m(r)|^2 , \tag{5.34}$$

whence

$$\|u\|^2_{r,r'} \le (r'-r) \exp\{2(r'-r)\mu\}\ \|u(\cdot,r)\|^2 . \tag{5.35}$$

To complete the proof of (5.20) recall that by Lemma 4.1, $u \in L_2^{2,\ell oc}(\Omega_h)$. It follows by Sobolev's imbedding theorem [1, p. 32] that there exists a constant $C_2 = C_2(h,r)$ such that

$$\|u(\cdot,r)\|^2 \le C_2(\|u\|^2_{h,r} + \|D_y\ u\|^2_{h,r}) \le C_2(\|u\|^2_{h,r} + \|\nabla u\|^2_{h,r}) . \tag{5.36}$$

Combining (5.35) and (5.36) gives (5.20).

Proof of Theorem 4.4. It must be shown that there exists a constant $C = C(K,r,r')$ such that for all $(p,\zeta) \in K$ and all $u \in F_{p,\zeta,r}$

$$\|u\|^2_{o,r'} + \|\nabla u\|^2_{o,r'} + \|\Delta u\|^2_{o,r'} \le C^2(\|u\|^2_{o,r} + \|\nabla u\|^2_{o,r} + \|\Delta u\|^2_{o,r}) . \tag{5.37}$$

Clearly it will suffice to show that

$$\|u\|^2_{r,r'} + \|\nabla u\|^2_{r,r'} + \|\Delta u\|^2_{r,r'} \le C^2(\|u\|^2_{o,r} + \|\nabla u\|^2_{o,r} + \|\Delta u\|^2_{o,r}) \tag{5.38}$$

since (5.37) then follows with $C^2 + 1$ instead of C^2. Moreover, every $u \in F_{p,\zeta,r}$ satisfies $\Delta u = -\pi(\zeta)u$ in Ω_r. Hence it will suffice to show that

$$(5.39) \qquad \|u\|^2_{r,r'} + \|\nabla u\|^2_{r,r'} \le C^2(\|u\|^2_{o,r} + \|\nabla u\|^2_{o,r} + \|\Delta u\|^2_{o,r})$$

since (5.38) then follows with C^2 Max $\{|\pi_p(\zeta)| + 1 : (p,\zeta) \in K\}$ instead of C^2.

To prove (5.39) note that the Fourier series argument used in the proof of Lemma 5.1 implies that (cf. (5.35))

$$(5.40) \qquad \|\nabla u\|^2_{r,r'} \le (r' - r) \exp\{2(r' - r)\mu\} \|\nabla u(\cdot,r)\|^2 .$$

Moreover, if $r'' = \frac{1}{2}(h + r)$ then $h < r'' < r$ and Sobolev's imbedding theorem implies that there exists a constant $C_3 = C_3(h,r)$ such that

$$(5.41) \qquad \|\nabla u(\cdot,r)\|^2 \le C_3 \|u\|^2_{2;r'',r}$$

where $\|\cdot\|_{2;r'',r}$ is the norm for $L_2^2(\Omega_{r'',r})$. Finally, the interior elliptic estimates of [1], applied to $v(x,y) = \exp\{-ipx\} u(x,y)$ and $L_p v = \Delta v + 2ip D_x v - p^2 v$ in $\Omega^\gamma_{h,r'}$, imply that there exists a constant $C_4 = C_4(h,r,r')$ such that

$$(5.42) \qquad \|u\|^2_{2;r'',r} \le C_4 (\|u\|^2_{h,r'} + \|\Delta u\|^2_{h,r'}) .$$

Moreover, since $\Delta u = -\pi(\zeta)u$ in $\Omega_{r,r'}$

$$(5.43) \qquad \begin{aligned} \|u\|^2_{h,r'} + \|\Delta u\|^2_{h,r'} &= \|u\|^2_{h,r} + \|u\|^2_{r,r'} + \|\Delta u\|^2_{h,r} + |\pi_p(\zeta)|^2 \|u\|^2_{r,r'} \\ &\le \|u\|^2_{h,r} + \|\Delta u\|^2_{h,r} + C_5(K) \|u\|^2_{r,r'} \end{aligned}$$

where $C_5(K) = \text{Max}\{|\pi_p(\zeta)|^2 : (p,\zeta) \in K\}$. Combining (5.20), (5.40), (5.41), (5.42) and (5.43) gives

$$(5.44) \quad \|u\|^2_{r,r'} + \|\nabla u\|^2_{r,r'} \le C_6(\|u\|^2_{h,r} + \|\nabla u\|^2_{h,r} + \|\Delta u\|^2_{h,r}) + C_7 \|u\|^2_{r,r'}$$

where $C_6 = \text{Max}(C_1, (r' - r) \exp\{2(r' - r)\mu(K)\} C_3 C_4)$ and $C_7 = (r' - r) \exp\{2(r' - r)\mu(K)\} C_3 C_4 C_5$. Finally, combining (5.20) and (5.44) gives (5.39) with $C^2 = \text{Max}(C_6, C_1 C_7)$.

It is worth remarking that an indirect (non-constructive) proof of Theorem 4.4 can be given by a compactness argument; see [30, Lemma 4.6] and Alber [3, Lemma 5.3].

<u>The Sesquilinear Form $A_{p,\zeta,r}$</u> in $L_2(\Omega_{0,r})$. Kato's first representation theorem [13, p. 322] associates a unique m-sectorial operator in $L_2(\Omega_{0,r})$ with each densely defined, closed, sectorial sesquilinear form in $L_2(\Omega_{0,r})$. Theorem 4.5 will be proved by constructing such a form $A_{p,\zeta,r}$ in $L_2(\Omega_{0,r})$ and showing that $A_{p,\zeta,r}$ is the associated m-sectorial operator. To motivate the definition of $A_{p,\zeta,r}$ note that if $v \in D(A_{p,\zeta,r})$ then application of Green's theorem gives

$$(v, A_{p,\zeta,r} v)_{0,r} = \|\nabla v\|^2_{0,r} - \int_{-\pi}^{\pi} \bar{v}\, D_y v \Big|_{y=r} dx \; . \tag{5.45}$$

The formal correctness of this equation is obvious. A rigorous proof based on the definition of $A_p^{\ell oc}$ is given below; see (5.115). Now $v = P_{p,\zeta,r} u$ where $u \in F_{p,\zeta,r}$ and u and v have Fourier expansions (4.7) for $h \le y < \infty$ and $h \le y \le r$, respectively. Moreover, Lemma 4.1 and the Sobolev theorems [1] imply that $u_m \in C^1[h,\infty)$, $v_m \in C^1[h,r]$, $u_m(y) = v_m(y)$ for $h \le y \le r$ and

$$u_m(y) = c_m \exp\{i y\, w_{p+m}(\zeta)\} \; , \; y \ge r \; . \tag{5.46}$$

Application of Parseval's formula to the integral in (5.45) gives the alternative representation

$$(v, A_{p,\zeta,r} v)_{0,r} = \|\nabla v\|^2_{0,r} - 2\pi i \sum_{m\in Z} w_{p+m}(\zeta)\; |v_m(r)|^2 \; . \tag{5.47}$$

The right-hand side of (5.47) will be used to define the form $A_{p,\zeta,r}$. Two cases, corresponding to the Dirichlet and Neumann boundary conditions respectively, must be distinguished. To this end define

$$\begin{aligned} G^D_{p,\zeta,r} &= L_2^{D,p,\ell oc}(\Omega) \,\cap \\ &\cap \{u \mid \operatorname{supp}\, (\Delta + \pi(\zeta))u \subset \Omega_{0,r};\ (4.28) \text{ holds in } L_2^{1,\ell oc}(\Omega_r)\} \; , \end{aligned} \tag{5.48}$$

$$\begin{aligned} G^N_{p,\zeta,r} &= L_2^{1,p,\ell oc}(\Omega_{0,r}) \,\cap \\ &\cap \{u \mid \operatorname{supp}\, (\Delta + \pi(\zeta))u \subset \Omega_{0,r};\ (4.28) \text{ holds in } L_2^{1,\ell oc}(\Omega_r)\} \; . \end{aligned} \tag{5.49}$$

The condensed notation $G_{p,\zeta,r}$ will be used to denote $G^D_{p,\zeta,r}$ or $G^N_{p,\zeta,r}$ in statements that hold for both. It is easy to verify that $G_{p,\zeta,R}$ is a Fréchet subspace of $L_2^{1,\ell oc}(\Omega)$. The notation $Q_{p,\zeta,r} : G_{p,\zeta,r} \to L_2(\Omega_{0,r})$ will be used for the natural projection defined by

$$(5.50)\qquad Q_{p,\zeta,r}u = u\Big|_{\Omega_{0,r}} \quad \text{for all } u \in G_{p,\zeta,r} .$$

The sesquilinear form $A_{p,\zeta,r}$ $(= A^D_{p,\zeta,r}$ or $A^N_{p,\zeta,r})$ and corresponding quadratic form are defined by

$$(5.51)\qquad D(A_{p,\zeta,r}) = Q_{p,\zeta,r}\, G_{p,\zeta,r} \subset L_2(\Omega_{0,r}) ,$$

$$(5.52)\qquad A_{p,\zeta,r}(v,v') = (\nabla v,\nabla v')_{0,r} - 2\pi i \sum_{m\in Z} w_{p+m}(\zeta)\, \overline{v_m(r)}\, v'_m(r)$$

for all $v,v' \in D(A_{p,\zeta,r})$, and

$$(5.53)\qquad A_{p,\zeta,r}(v) = A_{p,\zeta,r}(v,v) , \quad v \in D(A_{p,\zeta,r}) ,$$

and one has

<u>Theorem 5.2.</u> $A_{p,\zeta,r}$ is a densely defined, sectorial, closed sesquilinear form in $L_2(\Omega_{0,r})$.

The proof of this result requires a number of estimates which will be developed in a series of lemmas. The first lemma shows that (5.52) does indeed define a sesquilinear form on $L_2(\Omega_{0,r})$.

<u>Lemma 5.3.</u> For all $v,v' \in D(A_{p,\zeta,r})$ the series in (5.52) converges absolutely.

<u>Proof of Lemma 5.3.</u> It follows from Schwarz's inequality that it will suffice to prove that

$$(5.54)\qquad \sum_{m\in Z} w_{p+m}(\zeta)\ |v_m(r)|^2$$

converges absolutely when $v \in D(A_{p,\zeta,r})$ and

$$(5.55)\qquad v_m(r) = \frac{1}{2\pi}\int_{-\pi}^{\pi} e^{-i(p+m)x}\, v(x,r)\, dx .$$

To this end write $v = Q_{p,\zeta,r}u$ where $u \in G_{p,\zeta,r}$ and decompose u into

$$(5.56)\qquad u(X) = u'(X) + u''(X) , \quad X \in \Omega_r ,$$

as in the proof of Lemma 5.1.

Consider first the component u'. Parseval's relation implies that

$$(5.57)\qquad \int_{-\pi}^{\pi} |u'(x,y)|^2\, dx = 2\pi\, \Sigma'\, |c_m|^2 \exp\{-2y \operatorname{Im} w_{p+m}(\zeta)\}$$

$$\int_{-\pi}^{\pi} |\nabla u'(x,y)|^2 \, dx = 2\pi \, \Sigma' \, |c_m|^2 (|p+m|^2 + |w_{p+m}(\zeta)|^2) \exp\{-2y \operatorname{Im} w_{p+m}(\zeta)\}$$

(5.57 cont.)

for all $y \geq r$. Moreover, these are monotone decreasing functions of y that tend to zero exponentially at ∞. Hence $u' \in L_2^1(\Omega_r)$.

Next let $n, n' \in Z$ satisfy $n < n'$ and define

$$u'_{n,n'}(X) = \sum_{n}^{n'}{}' \, c_m \exp\{i\, x(p+m) + i\, y\, w_{p+m}(\zeta)\} \tag{5.58}$$

where $\sum_{n}^{n'}{}'$ denotes summation over the index set $\{m \mid \operatorname{Im} w_{p+m}(\zeta) > 0$ and $n \leq m \leq n'\}$. Applying Green's theorem to $u'_{n,n'}$ and $\overline{u'_{n,n'}}$ in $\Omega_{r,r'}$ gives

$$\int_{\Omega_{r,r'}} \{\overline{u'_{n,n'}} \, \Delta u'_{n,n'} + |\nabla u'_{n,n'}|^2\} \, dX = \int_{\partial\Omega_{r,r'}} \overline{u'_{n,n'}} \, D_\nu \, u'_{n,n'} \, ds \tag{5.59}$$

whence, using the Helmholtz equation and p-periodic boundary condition for $u'_{n,n'}$, one has

$$\|\nabla u'_{n,n'}\|^2_{r,r'} - \pi(\zeta) \, \|u'_{n,n'}\|^2_{r,r'} = \int_{-\pi}^{\pi} [\overline{u'_{n,n'}} \, D_y \, u'_{n,n'}]_r^{r'} \, dx \, . \tag{5.60}$$

Making $r' \to \infty$ and writing $\|\cdot\|_r = \|\cdot\|_{r,\infty}$ gives

$$\begin{aligned} \|\nabla u'_{n,n'}\|^2_r - \pi(\zeta) \, \|u'_{n,n'}\|^2_r &= -\int_{-\pi}^{\pi} \overline{u'_{n,n'}} \, D_y \, u'_{n,n'} \Big|_{y=r} dx \\ &= -2\pi i \sum_{n}^{n'}{}' \, w_{p+m}(\zeta) \, |u_m(r)|^2 \end{aligned} \tag{5.61}$$

where $v_m(r) = u_m(r) = c_m \exp\{i\, r\, w_{p+m}(\zeta)\}$. In particular, taking the real part of (5.61) gives

$$2\pi \sum_{n}^{n'}{}' \operatorname{Im} w_{p+m}(\zeta) \, |v_m(r)|^2 = \|\nabla u'_{n,n'}\|^2_r - \operatorname{Re} \pi(\zeta) \, \|u'_{n,n'}\|^2_r \, . \tag{5.62}$$

Hence the convergence of the Fourier series for u' in $L_2^1(\Omega_r)$ implies that

$$\Sigma' \operatorname{Im} w_{p+m}(\zeta) \, |v_m(r)|^2 < \infty \, . \tag{5.63}$$

The convergence is absolute because all the terms are non-negative.

Now consider the set

$$\{-i\, w_{p+m}(\zeta) \mid \operatorname{Im} w_{p+m}(\zeta) > 0\} \, . \tag{5.64}$$

Each member of the set satisfies

(5.65) $$|\arg(-i\, w_{p+m}(\zeta))| < \pi/2 \; .$$

Moreover, elements of the set (5.64) satisfy

(5.66) $$w_{p+m}(\zeta) \sim i\,|p+m| \,, \quad |m| \to \infty \,,$$

whence

(5.67) $$\arg(-i\, w_{p+m}(\zeta)) \to 0 \text{ when } |m| \to \infty \; .$$

It follows that

(5.68) $$\theta = \text{Max}\,\{|\arg(-i\, w_{p+m}(\zeta))| \,:\, \text{Im}\, w_{p+m}(\zeta) > 0\} < \pi/2 \; .$$

Hence if $\text{Im}\, w_{p+m}(\zeta) > 0$ then

(5.69) $$|\text{Re}\, w_{p+m}(\zeta)|\; |v_m(r)|^2 \le \tan\theta \;\text{Im}\, w_{p+m}(\zeta)\; |v_m(r)|^2$$

and (5.63) implies that

(5.70) $$\Sigma'\; |\text{Re}\, w_{p+m}(\zeta)|\; |v_m(r)|^2 < \infty \; .$$

(5.63), (5.70) and the finiteness of the sum defining u" imply the absolute convergence of the series (5.54).

Lemma 5.4. For each compact $K \subset M$ and each $r > h$ there exists an $a = a(K,r)$ such that for all $v \in \cup_{(p,\zeta)\in K} D(A_{p,\zeta,r})$ one has

(5.71) $$|2\pi\, \Sigma''\, w_{p+m}(\zeta)\; |v_m(r)|^2| \le \frac{1}{2}\, \|\nabla v\|^2_{0,r} + a\, \|v\|^2_{0,r} \; .$$

Proof of Lemma 5.4. Schwarz's inequality and the definition (5.55) imply that $2\pi\,|v_m(r)|^2 \le \|v(\cdot,r)\|^2_{L_2(-\pi,\pi)}$. Since $v \in L^1_2(\Omega_{0,r})$ it follows by Sobolev's imbedding theorem [1] that there exists a constant $C_0 = C_0(r)$ such that for all $\varepsilon \ge 1$ one has

(5.72) $$2\pi\, |v_m(r)|^2 \le C_0\, \varepsilon^{-1}(\|\nabla v\|^2_{0,r} + \varepsilon^2\, \|v\|^2_{0,r}) \; .$$

Next, note that if M(K) is the index set defined by (5.29) then M(K) is finite and hence

(5.73) $$C_1 = C_1(K) = \text{Max}\,\{|w_{p+m}(\zeta)| \,:\, (p,\zeta) \in K \text{ and } m \in M(K)\}$$

is finite for every compact $K \subset M$. Combining (5.72) and (5.73) gives

$$(5.74)\quad \begin{aligned} |2\pi\, \Sigma''\, w_{p+m}(\zeta)\, |v_m(r)|^2| &\le 2\pi\, C_1\, \Sigma''\, |v_m(r)|^2 \\ &\le 2\pi\, C_1 \sum_{m\in M} |v_m(r)|^2 \\ &\le C_0\, C_1\, M\, \varepsilon^{-1}(\|\nabla v\|^2_{0,r} + \varepsilon^2 \|v\|^2_{0,r}) \\ &\le \frac{1}{2}\|\nabla v\|^2_{0,r} + a\, \|v\|^2_{0,r} \end{aligned}$$

provided that $\varepsilon = \varepsilon(K,r) \ge 1$ is chosen such that $C_0\ C_1\ M\, \varepsilon^{-1} \le 1/2$ and $a = a(K,r)$ satisfies $a \ge C_0\ C_1\ M\, \varepsilon$.

Corollary 5.5. The sesquilinear form $A_{p,\zeta,r}$ is sectorial for all $(p,\zeta) \in M$. In fact, for each compact $K \subset M$ there exist constants $\gamma = \gamma(K) \in R$ and $\theta = \theta(K) < \pi/2$ such that for all $(p,\zeta) \in K$ and all $v \in D(A_{p,\zeta,r})$ with $\|v\|_{0,r} = 1$ one has

$$(5.75)\qquad A_{p,\zeta,r}(v) \in \{z \in C : |\arg\,(z-\gamma)| \le \theta\}\ .$$

Proof of Corollary 5.5. The proof generalizes one of Alber [3, Lemma 6.3]. Let $(p,\zeta) \in K$, $v \in D(A_{p,\zeta,r})$, $\|v\|_{0,r} = 1$ and write $A_{p,\zeta,r}(v) = I_1 + I_2$ where

$$(5.76)\qquad I_1 = \|\nabla v\|^2_{0,r} - 2\pi i\, \Sigma''\, w_{p+m}(\zeta)\, |v_m(r)|^2\ ,$$

and

$$(5.77)\qquad I_2 = -2\pi i\, \Sigma'\, w_{p+m}(\zeta)\, |v_m(r)|^2\ .$$

Then by Lemma 5.4 one has

$$(5.78)\qquad |\mathrm{Im}\ I_1| \le \frac{1}{2}\|\nabla v\|^2_{0,r} + a\ .$$

Similarly, the real part of I_1 satisfies

$$(5.79)\qquad \mathrm{Re}\ I \ge \|\nabla v\|^2_{0,r} - \frac{1}{2}\|\nabla v\|^2_{0,r} - a = \frac{1}{2}\|\nabla v\|^2_{0,r} - a\ .$$

Combining (5.78) and (5.79) gives $|\mathrm{Im}\ I_1| \le \mathrm{Re}\ I_1 + 2a$ whence

$$(5.80)\qquad I_1 \in \{z \in C : |\arg\,(z + 2a)| \le \pi/4\}\ .$$

Next, recall that $|\arg I_2| \leq \theta < \pi/2$ where $\theta = \theta(p,\zeta)$ is defined by (5.68). In fact, it is elementary to show that the limit relations (5.66), (5.67) hold uniformly for $(p,\zeta) \in K$ and hence there exists a $\theta_1 = \theta_1(K) < \pi/2$ such that $|\arg I_2| \leq \theta_1$ for all $(p,\zeta) \in K$. Combining this estimate with (5.80) gives (5.75) with $\gamma = -2a$ and $\theta = \text{Max}(\pi/4,\theta_1)$.

The proof that the form $A_{p,\zeta,r}$ is closed is based on the following generalization of an estimate of Alber [3, p. 269].

<u>Theorem 5.6</u>. For each $(p,\zeta) \in M$ and each $r' > r > h$ there exists a constant $C = C(p,\zeta,r,r')$ such that for all $v = Q_{p,\zeta,r}u$ with $u \in G_{p,\zeta,r}$ one has (see (4.30), (4.31) for notation)

(5.81) $$\|u\|^2_{1;0,r'} \leq C(|A_{p,\zeta,r}(v)| + \|v\|^2_{0,r}) .$$

The proof of Theorem 5.6 will be based on a number of related estimates which will be developed in a series of subsidiary lemmas. The first is

<u>Lemma 5.7</u>. Under the hypotheses of Theorem 5.6 one has

(5.82) $$\|\nabla u'\|^2_r - \pi(\zeta)\ \|u'\|^2_r = -2\pi i\ \Sigma'\ w_{p+m}(\zeta)\ |u_m(r)|^2$$

where u' by (5.22) and $u_m(y) = c_m \exp\{i y w_{p+m}(\zeta)\}$.

<u>Proof of Lemma 5.7</u>. The finiteness of the norms in (5.82) has already been noted; see (5.57). Passage to the limit $n \to -\infty$, $n' \to \infty$ in (5.61) gives (5.82).

<u>Lemma 5.8</u>. Under the hypotheses of Theorem 5.6 there exists a constant $C_1 = C_1(p,\zeta,r,r')$ such that for all $u \in G_{p,\zeta,r}$ one has

(5.83) $$\|u\|^2_{r,r'} \leq C_1\ \|u(\cdot,r)\|^2 .$$

<u>Proof of Lemma 5.8</u>. (5.83) follows from the proof of Lemma 5.1, inequality (5.35).

<u>Lemma 5.9</u>. Under the hypotheses of Theorem 5.6 there exists a constant $C_2 = C_2(p,\zeta,r,r')$ such that for all $u \in G_{p,\zeta,r}$ one has

(5.84) $$\|\nabla u''\|^2_{r,r'} \leq C_2\ \|u(\cdot,r)\|^2 .$$

The proof of Lemma 5.9, starting from (5.57), is exactly like that of Lemma 5.8 and is therefore omitted.

Lemma 5.10. Under the hypotheses of Theorem 5.6 there exists a constant $C_3 = C_3(p,\zeta,r,r')$ such that all $u \in G_{p,\zeta,r}$ one has

$$|\Sigma'' \, w_{p+m}(\zeta) \, |u_m(r)|^2| \le C_3 \, \|u(\cdot,r)\|^2 . \tag{5.85}$$

Proof of Lemma 5.10. One may take $C_3 = \text{Max}\,\{|w_{p+m}(\zeta)| : \text{Im}\, w_{p+m}(\zeta) \le 0\}$ and use (5.34).

Lemma 5.11. Under the hypotheses of Theorem 5.6 there exists a constant $C_4 = C_4(p,\zeta,r,r')$ such that one has

$$\|u'\|_r^2 \le C_4 \, \|u(\cdot,r)\|^2 . \tag{5.86}$$

Proof of Lemma 5.11. Integration of (5.57) over $r \le y < \infty$ gives

$$\|u'\|_r^2 = 2\pi \, \Sigma' \, |c_m|^2 \exp\{-2r \,\text{Im}\, w_{p+m}(\zeta)\}/2\,\text{Im}\, w_{p+m}(\zeta) \tag{5.87}$$

which with (5.34) implies (5.86) with $C_4(p,\zeta,r,r')$ defined by $C_4^{-1} = \text{Min}\,\{2\,\text{Im}\, w_{p+m}(\zeta) : \text{Im}\, w_{p+m}(\zeta) > 0\}$. This minimum is positive because $\text{Im}\, w_{p+m}(\zeta) \sim |p+m|$, $|m| \to \infty$ (see (5.66)).

Lemma 5.12. Under the hypotheses of Theorem 5.6 to each $\alpha > 0$ there corresponds a constant $\theta_\alpha = \theta_\alpha(h,r)$ such that for all $u \in G_{p,\zeta,r}$ one has

$$\|u(\cdot,r)\|^2 \le \alpha \, \|\nabla u\|_{0,r}^2 + \theta_\alpha \, \|u\|_{0,r}^2 . \tag{5.88}$$

Proof of Lemma 5.12. Recall that $u \in C^\infty(\overline{\Omega}_r)$ and hence $u_m(y) \in C^\infty[r,\infty)$. Hence by a Sobolev inequality there is a constant $\gamma = \gamma(h,r)$ such that for all $\varepsilon \ge 1$ one has [1]

$$|u_m(r)|^2 \le \gamma \, \varepsilon^{-1} \left(\int_h^r |u_m(y)|^2 \, dy + \varepsilon^2 \int_h^r |u_m'(y)|^2 \, dy \right) . \tag{5.89}$$

Moreover, by Parseval's relation,

$$\|u(\cdot,y)\|^2 = \int_{-\pi}^{\pi} |u(x,y)|^2 \, dx = 2\pi \sum_{m\in Z} |u_m(y)|^2 , \quad y \ge h , \tag{5.90}$$

whence

$$(5.91)\qquad \|u\|_{h,r}^2 = 2\pi \sum_{m\in Z} \int_h^r |u_m(y)|^2\, dy ,$$

$$(5.92)\qquad \|D_y u\|_{h,r}^2 = 2\pi \sum_{m\in Z} \int_h^r |D_y\, u_m(y)|^2\, dy .$$

Combining (5.89)-(5.92) gives the estimate

$$(5.93)\qquad \begin{aligned} \|u(\cdot,r)\|^2 &\le \gamma\, \varepsilon^{-1}(\|D_y u\|_{h,r}^2 + \varepsilon^2\, \|u\|_{h,r}^2) \\ &\le \gamma\, \varepsilon^{-1}(\|\nabla u\|_{0,r}^2 + \varepsilon^2\, \|u\|_{0,r}^2) . \end{aligned}$$

Choosing $\gamma(h,r)\varepsilon^{-1} = \alpha$, $\gamma(h,r)\varepsilon = \gamma^2(h,r)/\alpha = \theta_\alpha(h,r)$ in (5.93) gives (5.88).

Proof of Theorem 5.6. The definition (5.52), (5.53) implies that for all $v = Q_{p,\zeta,r}u \in D(A_{p,\zeta,r})$

$$(5.94)\qquad A_{p,\zeta,r}(v) = \|\nabla v\|_{0,r}^2 - 2\pi i \sum_{m\in Z} w_{p+m}(\zeta)\, |u_m(r)|^2 .$$

Combining this with Lemma 5.7 gives the representation

$$(5.95)\qquad A_{p,\zeta,r}(v) = \|\nabla v\|_{0,r}^2 + \|\nabla u'\|_r^2 - \pi(\zeta)\, \|u'\|_r^2 - 2\pi i\, \Sigma''\, w_{p+m}(\zeta)\, |u_m(r)|^2$$

whence

$$(5.96)\qquad \begin{aligned} \|\nabla v\|_{0,r}^2 + \|\nabla u'\|_{r,r'}^2 &\le \|\nabla v\|_{0,r}^2 + \|\nabla u'\|_r^2 \\ &= \mathrm{Re}\, \{A_{p,\zeta,r}(v) + \pi(\zeta)\, \|u'\|_r^2 + 2\pi i\, \Sigma''\, w_{p+m}(\zeta)\, |u_m(r)|^2\} \\ &\le |A_{p,\zeta,r}(v)| + |\pi(\zeta)|\, \|u'\|_r^2 + 2\pi\, |\Sigma''\, w_{p+m}(\zeta)\, |u_m(r)|^2| . \end{aligned}$$

It follows that

$$(5.97)\qquad \begin{aligned} \|u\|_{1;0,r'}^2 &= \|\nabla v\|_{0,r}^2 + \|\nabla u'\|_{r,r'}^2 + \|\nabla u''\|_{r,r'}^2 + \|v\|_{0,r}^2 + \|u\|_{r,r'}^2 \\ &\le |A_{p,\zeta,r}(v)| + |\pi(\zeta)|\, \|u'\|_r^2 + 2\pi\, |\Sigma''\, w_{p+m}(\zeta)\, |u_m(r)|^2| \\ &\quad + \|\nabla u''\|_{r,r'}^2 + \|u\|_{r,r'}^2 + \|v\|_{0,r}^2 . \end{aligned}$$

Combining (5.97) and the estimates of Lemmas 5.8-5.11 gives

$$\|u\|^2_{1;0,r'} \le |A_{p,\zeta,r}(v)| + C_5 \ \|u(\cdot,r)\|^2 + \|v\|^2_{0,r} \tag{5.98}$$

where

$$C_5 = C_5(p,\zeta,r,r') = C_1 + C_2 + 2\pi\ C_3 + |\pi(\zeta)|\ C_4 \ . \tag{5.99}$$

On combining (5.98) and (5.88), and recalling that $u = v$ in $\Omega_{0,r}$, one finds

$$\|u\|^2_{1;0,r'} \le |A_{p,\zeta,r}(v)| + (C_5\alpha)\ \|\nabla v\|^2_{0,r} + (C_6\ \theta_\alpha + 1)\ \|v\|^2_{0,r} \tag{5.100}$$

where $\alpha > 0$ is arbitrary. Defining α by $C_5\alpha = 1/2$ and $C = C(p,\zeta,r,r')$ $= 2(C_5\theta_\alpha + 1)$ gives

$$\|u\|^2_{1;0,r'} \le \frac{1}{2}\ C(|A_{p,\zeta,r}(v)| + \|v\|^2_{0,r}) + \frac{1}{2}\ \|\nabla v\|^2_{0,r} \tag{5.101}$$

since $\frac{1}{2}\ C = C_5\alpha + 1 \ge 1$. Finally, (5.101) implies (5.81) because $\|\nabla v\|_{0,r} \le \|u\|_{1;0,r'}$.

<u>Proof of Theorem 5.2</u>. The denseness of $D(A_{p,\zeta,r})$ in $L_2(\Omega_{0,r})$ follows from the obvious inclusion $C_0^\infty(\Omega_{0,r}) = Q_{p,\zeta,r}\ C_0^\infty(\Omega_{0,r}) \subset D(A_{p,\zeta,r})$. The sectorial property of $A_{p,\zeta,r}$ was proved as Corollary 5.5 above. To prove that $A_{p,\zeta,r}$ is closed let $v^{(n)} = Q_{p,\zeta,r}\ u^{(n)}$, with $u^{(n)} \in G_{p,\zeta,r}$, be $A_{p,\zeta,r}$-convergent to $v \in L_2(\Omega_{0,r})$; i.e., $v^{(n)} \to v$ in $L_2(\Omega_{0,r})$ and $A_{p,\zeta,r}(v^{(n)} - v^{(m)}) \to 0$ when $n,m \to \infty$. It must be shown that $v = Q_{p,\zeta,r}u$ where $u \in G_{p,\zeta,r}$ and $A_{p,\zeta,r}(v - v^{(n)}) \to 0$ when $n \to \infty$ [13, p. 313]. Now Theorem 5.6 applied to $v^{(n)} - v^{(m)} = Q_{p,\zeta,r}(u^{(n)} - u^{(m)})$ implies that $\{u^{(n)}\}$ is a Cauchy sequence in $G_{p,\zeta,r}$ and hence $\lim u^{(n)} = u \in G_{p,\zeta,r}$ exists. Clearly, $v = Q_{p,\zeta,r}u$ since $Q_{p,\zeta,r}$ is bounded. Moreover, the convergence of $\{u^{(n)}\}$ to u in $G_{p,\zeta,r}$ implies that $\|\nabla v - \nabla v^{(n)}\|_{0,r} \to 0$ when $n \to \infty$. Hence, the representation (5.94) of $A_{p,\zeta,r}(v)$ implies that to complete the proof of Theorem 5.2 it will be enough to show that

$$\lim_{n\to\infty} \sum_{m\in Z} w_{p+m}(\zeta)\ |u_m(r) - u_m^{(n)}(r)|^2 = 0\ . \tag{5.102}$$

Now Lemma 5.3 and the relation (5.60), applied to the partial sums of the Fourier series of $u \in G_{p,\zeta,r}$ in $\Omega_{r,r'}$ imply that

$$2\pi i \sum_{m\in Z} w_{p+m}(\zeta)\ |u_m(r)|^2 = (u(\cdot,r'),D_y u(\cdot,r'))_{L_2(-\pi,\pi)} + \pi(\zeta)\ \|u\|^2_{r,r'} - \|\nabla u\|^2_{r,r'} \tag{5.103}$$

It follows that (5.102) holds if

$$(5.104)\quad \lim_{n\to\infty} (u(\cdot,r') - u^{(n)}(\cdot,r'), D_y u(\cdot,r') - D_y u^{(n)}(\cdot,r'))_{L_2(-\pi,\pi)} = 0 .$$

To prove this define $s = 1/2\ (r + r')$, $s' = r + r'$ so that $r < s < r' < s'$. Then a Sobolev imbedding theorem [1] implies that

$$(5.105)\quad \begin{aligned} |(u(\cdot,r'), D_y u(\cdot,r'))| &\le \|u(\cdot,r')\|\ \|D_y u(\cdot,r')\| \\ &\le \|u\|_{1;s,r'}\ \|u\|_{2;s,r'} \\ &\le \|u\|^2_{2;s,r'} . \end{aligned}$$

Moreover, the interior elliptic estimates of [1] imply that there exists a $C = C(r,r')$ such that (see (5.42))

$$(5.106)\quad \|u\|^2_{2;s,r'} \le C(\|u\|^2_{r,s'} + \|\Delta u\|^2_{r,s'}) .$$

Since $\Delta u = -\pi(\zeta)u$ in Ω_r, (5.105) and (5.106) imply

$$(5.107)\quad |(u(\cdot,r'), D_y u(\cdot,r'))| \le C'\ \|u\|^2_{r,s'}$$

where $C' = C'(p,\zeta,r,r')$. Applying (5.107) to $u - u^{(n)}$ gives (5.104). This completes the proof of Theorem 5.2. Note that the proof actually implies

<u>Corollary 5.13</u>. $Q_{p,\zeta,r}$ is a topological isomorphism of the Fréchet space $G_{p,\zeta,r}$ onto $D(A_{p,\zeta,r}) = Q_{p,\zeta,r}\, G_{p,\zeta,r}$, topologized by the norm

$$(5.108)\quad (|A_{p,\zeta,r}(v)| + \|v\|^2_{0,r})^{1/2} .$$

<u>Proof of Theorem 4.5</u>. The densely defined, sectorial, closed sesquilinear form $A_{p,\zeta,r}$ is associated with a unique m-sectorial operator $T_{p,\zeta,r}$ in $L_2(\Omega_{0,r})$ by Kato's first representation theorem [13, p. 322]. Theorem 4.5 will be proved by showing that $A_{p,\zeta,r} = T_{p,\zeta,r}$.

<u>The Inclusion $A_{p,\zeta,r} \subset T_{p,\zeta,r}$</u>. To prove this let $v \in D(A_{p,\zeta,r}) = P_{p,\zeta,r}\, F_{p,\zeta,r} \subset D(A_{p,\zeta,r})$ and write $z = -\Delta v = A_{p,\zeta,r} v$. It will be shown that

$$(5.109)\quad A_{p,\zeta,r}(v',v) = (v', A_{p,\zeta,r} v)_{0,r} = (v',z)_{0,r} , \quad v' \in D(A_{p,\zeta,r}) .$$

Note that this implies that $v \in D(T_{p,\zeta,r})$ and $T_{p,\zeta,r}v = z = A_{p,\zeta,r}v$, whence $A_{p,\zeta,r} \subset T_{p,\zeta,r}$.

Equation (5.109) will be proved by applying the generalized Dirichlet or Neumann boundary condition to v; i.e., the integral identities of the definitions (3.19), (3.20) of $D(A_p^{\ell oc})$. To this end let $r' > r$ and let $\phi_{r,r'}(y) \in C^\infty(R)$ be a cut-off function with the properties

$$\phi_{r,r'}(y) = \begin{cases} 1 , & y \le (2r + r')/3 , \\ 0 , & y \ge (r + 2r')/3 , \end{cases} \tag{5.110}$$

and $\phi'_{r,r'}(y) \le 0$ (whence $0 \le \phi_{r,r'}(y) \le 1$). Then

$$v_{r,r'} = \phi_{r,r'}v' \in L_2^{D,p,com}(\Omega) \text{ or } L_2^{1,p,com}(\Omega) \tag{5.111}$$

and

$$\nabla v_{r,r'} = \phi_{r,r'} \nabla v' + \phi'_{r,r'} v'\hat{y} \tag{5.112}$$

where $\hat{y}$ is a unit vector in the y-direction. The integral identity of (3.19) or (3.20), applied to $v \in D(A_p^{\ell oc})$ and $v_{r,r'}$ gives

$$\begin{aligned} 0 &= (v_{r,r'},\Delta v)_{0,r'} + (\nabla v_{r,r'},\nabla v)_{0,r'} \\ &= (\phi_{r,r'}v',\Delta v)_{0,r'} + (\phi_{r,r'}\nabla v',\nabla v)_{0,r'} + (\phi'_{r,r'}v',D_y v)_{0,r'} . \end{aligned} \tag{5.113}$$

Now the last term satisfies

$$\begin{aligned} (\phi'_{r,r'}v',D_y v)_{0,r'} &= \int_r^{r'} \int_{-\pi}^{\pi} \phi'_{r,r'} \overline{v'(x,y)}\, D_y v(x,y)\, dxdy \\ &= \int_r^{r'} \phi'_{r,r'}(y) \left(\int_{-\pi}^{\pi} \overline{v'(x,y)}\, D_y v(x,y) dx\right) dy \\ &\to -\int_{-\pi}^{\pi} \overline{v'(x,\gamma)}\, D_y v(x,\gamma)\, dx \\ &= -2\pi i \sum_{m\in Z} w_{p+m}(\zeta)\, \overline{v'_m(r)}\, v_m(r), \quad r' \to r; \end{aligned} \tag{5.114}$$

see [30, p. 57] for a similar calculation. Thus passage to the limit $r' \to r$ in (5.113) gives

$$(v',\Delta v)_{0,r} + (\nabla v',\nabla v)_{0,r} - 2\pi i \sum_{m\in Z} w_{p+m}(\zeta)\, \overline{v'_m(r)}\, v_m(r) = 0 \tag{5.115}$$

for all $v' \in D(A_{p,\zeta,r})$. The definition (5.52) of $A_{p,\zeta,r}$ implies that (5.115) is equivalent to (5.109).

<u>The Inclusion</u> $T_{p,\zeta,r} \subset A_{p,\zeta,r}$. To prove this let $v \in D(T_{p,\zeta,r})$ and $T_{p,\zeta,r}v = z \in L_2(\Omega_{0,r})$. This is equivalent to the identity

$$A_{p,\zeta,r}(v',v) = (v',z)_{0,r} \tag{5.116}$$

or

$$(\nabla v',\nabla v)_{0,r} - 2\pi i \sum_{m\in Z} w_{p+m}(\zeta)\, \overline{v'_m(r)}\, v_m(r) = (v',z)_{0,r} \tag{5.117}$$

for all $v' \in D(A_{p,\zeta,r})$. Taking $v' \in C_0^\infty(\Omega_{0,r})$ gives

$$-\Delta v = z \text{ in } \Omega_{0,r} \tag{5.118}$$

by elementary distribution theory. Thus to complete the proof it is enough to show that $v \in D(A_{p,\zeta,r})$. Note that the definitions of $F_{p,\zeta,r}$ and $G_{p,\zeta,r}$ imply

$$F_{p,\zeta,r} = G_{p,\zeta,r} \cap L_2^{\ell oc}(\Delta,\Omega) . \tag{5.119}$$

Thus it will suffice to show that $u = Q_{p,\zeta,r}^{-1}v$ satisfies $\Delta u \in L_2^{\ell oc}(\Omega)$. This will be done by calculating the distribution Δu. To this end note that for all $\psi \in C_0^\infty(\Omega)$ one has

$$(-\Delta\psi,u)_{L_2(\Omega)} = (\nabla\psi,\nabla u)_{L_2(\Omega)} \tag{5.120}$$

because $u \in L_2^{1,\ell oc}(\Omega)$. Thus

$$(-\Delta\psi,u)_{L_2(\Omega)} = (\nabla\psi,\nabla v)_{0,r} + (\nabla\psi,\nabla u)_{r,\infty} . \tag{5.121}$$

Now equation (5.117) with $v' = \psi$ gives

$$(\nabla\psi,\nabla v)_{0,r} = (\psi,z)_{0,r} + 2\pi i \sum_{m\in Z} w_{p+m}(\zeta)\, \overline{\psi_m(r)}\, v_m(r) . \tag{5.122}$$

It will be shown that the last term in (5.121) satisfies

$$(\nabla\psi,\nabla u)_{r,\infty} = \pi(\zeta)(\psi,u)_{r,\infty} - 2\pi i \sum_{m\in Z} w_{p+m}(\zeta)\, \overline{\psi_m(r)}\, v_m(r) . \tag{5.123}$$

Adding equations (5.122) and (5.123) and using (5.121) gives

$$(5.124)\qquad (-\Delta\psi,u)_{L_2(\Omega)} = (\psi,z)_{0,r} + \pi(\zeta)(\psi,u)_{r,\infty} = (\psi,f)_{L_2(\Omega)}$$

where

$$(5.125)\qquad f(X) = \begin{cases} z(X)\ , & X \in \Omega_{0,r}, \\ \pi(\zeta)\ u(X)\ , & X \in \Omega_{r,\infty}\ . \end{cases}$$

Thus $-\Delta u = f \in L_2^{\ell oc}(\Omega)$.

The proof of Theorem 4.5 will be completed by verifying (5.123). To this end recall that $u \in G_{p,\zeta,r}$ and $\Delta u = -\pi(\zeta)u$ as a distribution in $\Omega_{r,\infty}$. Now define $\theta_{r,r'}(y) = 1 - \phi_{r,r'}(y)$, where $\phi_{r,r'}$ is defined as above, and define

$$(5.126)\qquad \psi_{r,r'} = \theta_{r,r'}\psi \in C_0^\infty(\Omega)\ .$$

Then the distribution definitions of ∇u and Δu in $\Omega_{r,\infty}$ imply

$$(5.127)\qquad \begin{aligned} (\nabla\psi_{r,r'},\nabla u)_{r,\infty} &= (-\Delta\psi_{r,r'},u)_{r,\infty} = (\psi_{r,r'},-\Delta u)_{r,\infty} \\ &= \pi(\zeta)(\psi_{r,r'},u)_{r,\infty}\ . \end{aligned}$$

On the other hand, proceeding as in the first part of the proof one finds

$$(5.128)\qquad \begin{aligned} (\nabla\psi_{r,r'},\nabla u)_{r,\infty} &= (\theta_{r,r'}\nabla\psi,\nabla u)_{r,\infty} + (\theta'_{r,r'}\psi,D_y u)_{r,\infty} \\ &\to (\nabla\psi,\nabla u)_{r,\infty} + 2\pi i \sum_{m\in Z} w_{p+m}(\zeta)\ \overline{\psi_m(r)}\ u_m(r) \end{aligned}$$

when $r' \to r$. Thus passage to the limit $r' \to r$ in (5.127) gives (5.123) because $v = Q_{p,\zeta,r}u$ satisfies $v_m(r) = u_m(r)$.

Proof of Theorem 4.6. The proof of the continuity of $\{A_{p,\zeta,r} \mid (p,\zeta) \in M\}$ will be based on a criterion established by Kato [13, Theorem IV-2.29]. Thus for each $(p_0,\zeta_0) \in M$ one must construct a Hilbert space $\mathcal{H}$, a neighborhood $N(p_0,\zeta_0) \subset M$, operators $U(p,\zeta)$, $V(p,\zeta) \in B(\mathcal{H},L_2(\Omega_{0,r}))$ for $(p,\zeta) \in N(p_0,\zeta_0)$, and operators $U,V \in B(\mathcal{H},L_2(\Omega_{0,r}))$ with the properties that $U(p,\zeta)$ and U map $\mathcal{H}$ one-to-one onto $D(A_{p,\zeta,r})$ and $D(A_{p_0,\zeta_0,r})$, respectively,

$$(5.129)\qquad A_{p,\zeta,r}\,U(p,\zeta) = V(p,\zeta),\ A_{p_0,\zeta_0,r}\,U = V\ ,$$

and

$$\|U(p,\zeta) - U\| \to 0 \ , \ \|V(p,\zeta) - V\| \to 0 \text{ when } (p,\zeta) \to (p_0,\zeta_0) \ . \tag{5.130}$$

The space $\mathcal{H}$ will be defined by

$$\mathcal{H} = D(A_{p_0,\zeta_0,r}) \subset L_2^1(\Delta,\Omega_{0,r}) \ . \tag{5.131}$$

Theorem 4.4 implies that $\mathcal{H}$ is closed in the topology of $L_2^1(\Delta,\Omega_{0,r})$ and hence is a Hilbert space. Next a neighborhood $N(p_0,\zeta_0)$ and linear operators

$$J(p,\zeta,p_0,\zeta_0) \in B(\mathcal{H},L_2^1(\Delta,\Omega_{0,r})) \ , \quad (p,\zeta) \in N(p_0,\zeta_0) \ , \tag{5.132}$$

will be constructed with the properties

(5.133) $J(p,\zeta,p_0,\zeta_0)$ maps $\mathcal{H}$ one-to-one onto $D(A_{p,\zeta,r})$,

(5.134) $J(p_0,\zeta_0,p_0,\zeta_0) = E$ is the natural embedding of $\mathcal{H}$ in $L_2^1(\Delta,\Omega_{0,r})$,

(5.135) $(p,\zeta) \to J(p,\zeta,p_0,\zeta_0) \in B(\mathcal{H},L_2^1(\Delta,\Omega_{0,r}))$ is continuous at (p_0,ζ_0) .

The desired operators can then be defined by

$$U(p,\zeta) = E_0 \ J(p,\zeta,p_0,\zeta_0) \ , \ U = U(p_0,\zeta_0) \ , \tag{5.136}$$

$$V(p,\zeta) = A_{p,\zeta,r} \ U(p,\zeta) \ , \quad V = V(p_0,\zeta_0) \ , \tag{5.137}$$

where $E_0 : L_2^1(\Delta,\Omega_{0,r}) \to L_2(\Omega_{0,r})$ is the natural embedding. It is clear that these operators are in $B(\mathcal{H},L_2(\Omega_{0,r}))$ and $U(p,\zeta)$, U map $\mathcal{H}$ one-to-one onto $D(A_{p,\zeta,r})$, $D(A_{p_0,\zeta_0,r})$, respectively. Equations (5.129) hold by definition. Moreover,

$$\begin{aligned} \|U(p,\zeta) - U\| &= \|E_0(J(p,\zeta,p_0,\zeta_0) - J(p_0,\zeta_0,p_0,\zeta_0))\| \\ &\le \|J(p,\zeta,p_0,\zeta_0) - E\| \to 0 \end{aligned} \tag{5.138}$$

when $(p,\zeta) \to (p_0,\zeta_0)$ by (5.134), (5.135). Similarly, for all $u \in \mathcal{H}$,

$$\begin{aligned}
\|(V(p,\zeta) - V)u\|_{0,r} &= \|\Delta J(p,\zeta,p_0,\zeta_0)u - \Delta J(p_0,\zeta_0,p_0,\zeta_0)u\|_{0,r} \\
&\le \|J(p,\zeta,p_0,\zeta_0)u - E\,u\|_{1,\Delta;0,r} \qquad (5.139) \\
&\le \|J(p,\zeta,p_0,\zeta_0) - E\| \; \|u\|_{\mathcal{H}}
\end{aligned}$$

whence

$$\|V(p,\zeta) - V\| \le \|J(p,\zeta,p_0,\zeta_0) - E\| \to 0 \qquad (5.140)$$

when $(p,\zeta) \to (p_0,\zeta_0)$. The proof of Theorem 4.6 will be completed by constructing the family $J(p,\zeta,p_0,\zeta_0)$. The cases of the Dirichlet and Neumann boundary conditions will be treated separately.

<u>Construction of J - The Dirichlet Case</u>. The construction generalizes one of Alber [3]. To describe it let $v \in \mathcal{H} = D(A_{p_0,\zeta_0,r})$; i.e., $v = P_{p,\zeta,r}u$, $u \in F_{p_0,\zeta_0,r}$. The Fourier expansions of v and u have the forms

$$v(x,y) = \sum_{m\in Z} v_m(y) \exp\{i(p_0 + m)x\} \;,\quad (x,y) \in \Omega_{h,r} \;, \qquad (5.141)$$

$$u(x,y) = \sum_{m\in Z} u_m(y) \exp\{i(p_0 + m)x\} \;,\quad (x,y) \in \Omega_{h,\infty} \;. \qquad (5.142)$$

Moreover, $v_m(y) = u_m(y)$ for $h \le y \le r$ and

$$u_m(y) = c_m \exp\{i\,y\,w_{p_0+m}(\zeta_0)\} \text{ for } y \ge r \;. \qquad (5.143)$$

Now introduce a function $\xi \in C^\infty(R)$ such that

$$\xi(y) = \begin{cases} 1 \text{ for } -\infty < y \le r_1 = (r + 2h)/3 \;, \\ 0 \text{ for } r_2 = (2r + h)/3 \le y < \infty \end{cases} \qquad (5.144)$$

and $\xi'(y) \le 0$ (whence $0 \le \xi(y) \le 1$), and define, for each $y \in R$,

$$d_m(p,\zeta,p_0,\zeta_0,y) = \exp\{i\,y\,[w_{p+m}(\zeta) - w_{p_0+m}(\zeta_0)]\}[1 - \xi(y)] + \xi(y) \;. \qquad (5.145)$$

<u>Choice of $N(p_0,\zeta_0)$</u>. The equicontinuity of the functions $w_{p+m}(\zeta)$, Theorem 4.2, implies that there exists a neighborhood $N(p_0,\zeta_0) \subset M$ such that

$$|\exp\{i\,y\,[w_{p+m}(\zeta) - w_{p_0+m}(\zeta_0)]\} - 1| < 1/2 \qquad (5.146)$$

for all $(p,\zeta) \in N(p_0,\zeta_0)$, $m \in Z$ and $y \in R$. Thus, using $||z_1| - |z_2|| \le |z_1 - z_2|$ one has

$$(5.147)\qquad \begin{aligned} ||d_m(p,\zeta,p_0,\zeta_0,y)| - 1| &\le |d_m(p,\zeta,p_0,\zeta_0,y) - 1| \\ &\le |\exp\{i\,y\,[w_{p+m}(\zeta) - w_{p_0+m}(\zeta_0)]\} - 1|\ |1-\xi(y)| < 1/2 \end{aligned}$$

and hence

$$(5.148)\qquad 1/2 < |d_m(p,\zeta,p_0,\zeta_0,y)| < 3/2$$

for all $(p,\zeta) \in N(p_0,\zeta_0)$, $m \in Z$ and $y \in R$.

Definition. For all $v \in \mathcal{H} = D(A_{p_0,\zeta_0,r})$ with expansion (5.141) on $\Omega_{h,r}$ let

$$(5.149)\qquad J(p,\zeta,p_0,\zeta_0)\, v(x,y) = \exp\{i(p-p_0)x\} \begin{cases} \sum_{m\in Z} d_m(p,\zeta,p_0,\zeta_0,y)\, v_m(y) \exp\{i(p_0+m)x\} & \text{in } \Omega_{h,r}, \\ v(x,y) & \text{in } \Omega_{0,h}. \end{cases}$$

Note that $d_m(p,\zeta,p_0,\zeta_0,y) \equiv 1$ and hence $J(p,\zeta,p_0,\zeta_0)\, v(x,y) = \exp\{i(p-p_0)x\}\, v(x,y)$ for $(x,y) \in \Omega_{h,r_1}$. Thus the definition produces no discontinuities at $y = h$. The proof that J has the properties (5.132) -(5.135) will be developed in several lemmas.

Lemma 5.14. There exists a constant $M = M(N(p_0,\zeta_0))$ such that

$$(5.150)\qquad |D_y^k\, d_m(p,\zeta,p_0,\zeta_0,y)| \le M$$

for all $(p,\zeta) \in N(p_0,\zeta_0)$, $m \in Z$, $y \in R$ and $k = 0,1,2$.

This result follows easily from the definition (5.145) and the equicontinuity of the family $\{w_{p+m}(\zeta)\}$.

Lemma 5.15. J satisfies (5.132); i.e., for all $v \in D(A_{p_0,\zeta_0,r})$ one has $J(p,\zeta,p_0,\zeta_0)v \in L_2^1(\Delta,\Omega_{0,r})$ and there exists a $C = C(p_0,\zeta_0)$ such that

$$(5.151)\qquad \|J(p,\zeta,p_0,\zeta_0)v\|_{1,\Delta;0,r} \le C\, \|v\|_{1,\Delta;0,r}$$

for all $v \in D(A_{p_0,\zeta_0,r})$ and all $(p,\zeta) \in N(p_0,\zeta_0)$.

Proof of Lemma 5.15. For all $v \in D(A_{p_0,\zeta_0,r})$ one has

$$(5.152)\qquad \begin{aligned} \|v\|^2_{1,\Delta;0,r} &= \int_{\Omega_{0,r}} \{|v|^2 + |\nabla v|^2 + |\Delta v|^2\}\, dX \\ &= \|v\|^2_{1,\Delta;0,h} + \|v\|^2_{1,\Delta;h,r} \\ &= \|v\|^2_{1,\Delta;0,h} + 2\pi \sum_{m\in Z} \int_h^r I^0_m(y)\, dy \end{aligned}$$

where

$$(5.153)\qquad I^0_m(y) = (1 + |p_0+m|^2)\, |v_m|^2 + |D_y v_m|^2 + |D^2_y v_m - (p_0+m)^2\, v_m|^2 .$$

Similarly, writing

$$(5.154)\qquad J\, v_m(y) = d_m(p,\zeta,p_0,\zeta_0,y)\, v_m(y) ,$$

one has

$$(5.155)\qquad \|J(p,\zeta,p_0,\zeta_0)v\|^2_{1,\Delta;0,r} = \|\exp\{i(p-p_0)\cdot\}v\|^2_{1,\Delta;0,h} + 2\pi \sum_{m\in Z} \int_h^r I_m(y)\, dy$$

where

$$(5.156)\qquad I_m(y) = (1+|p+m|^2)\, |Jv_m|^2 + |D_y Jv_m|^2 + |D^2_y Jv_m - (p+m)^2\, Jv_m|^2 .$$

Now a simple calculation gives the estimate

$$(5.157)\qquad \|\exp\{i(p-p_0)\cdot\}v\|_{1,\Delta;0,h} \le C_1\, \|v\|_{1,\Delta;0,h}$$

where $C_1 = C_1(N(p_0,\zeta_0))$. Similarly, Lemma 5.14 implies that there is a constant $C_2 = C_2(N(p_0,\zeta_0))$ such that

$$(5.158)\qquad I_m(y) \le C_2^2\, I^0_m(y)$$

for all $(p,\zeta) \in N(p_0,\zeta_0)$, $m \in Z$ and $h \le y < \infty$. It follows that $J(p,\zeta,p_0,\zeta_0)v \in L^1_2(\Delta,\Omega_{0,r})$ and (5.151) holds with $C^2 = \text{Max}\,(C_1^2, 2\pi C_2^2)$.

Lemma 5.16. For all $v \in D(A_{p_0,\zeta_0,r})$ one has

$$(5.159)\qquad J(p,\zeta,p_0,\zeta_0)v \in D(A_{p,\zeta,r}) .$$

Proof of Lemma 5.16. Since $D(A_{p,\zeta,r}) = P_{p,\zeta,r} F_{p,\zeta,r}$ it must be shown that $\tilde{v} = J(p,\zeta,p_0,\zeta_0)v$ has a continuation $\tilde{u}$ to Ω which is in $F_{p,\zeta,r}$. Recall that for $h \le y \le r$ one has $u_m(y) = v_m(y)$ and hence

$$\tilde{v}(x,y) = \sum_{m\in Z} d_m(p,\zeta,p_0,\zeta_0,y)\, u_m(y) \exp\{i(p+m)x\} \tag{5.160}$$

where $u_m(y)$ is defined by (5.142), (5.143). Moreover, for $h < r_2 \le y \le r$ one has

$$d_m(p,\zeta,p_0,\zeta_0,y) = \exp\{iy\,[w_{p+m}(\zeta) - w_{p_0+m}(\zeta_0)]\} \tag{5.161}$$

and hence it is natural to define the continuation of $\tilde{v}$ by (5.142), (5.143) and

$$\tilde{u}(x,y) = \sum_{m\in Z} c_m \exp\{ix(p+m) + iy\,w_{p+m}(\zeta)\}\ ,\ y \ge r\ . \tag{5.162}$$

It is clear from the convergence of (5.142) in $L_2^{2,\ell oc}(\Omega_h)$ (Lemma 4.1) and Lemma 5.14 that (5.162) converges in $L_2^{2,\ell oc}(\Omega_r)$ and hence $\tilde{u} \in L_2^{1,\ell oc}(\Delta,\Omega)$. Also, the p_0-periodic boundary condition satisfied by v, together with (5.149) and (5.162), imply that $\tilde{u}$ satisfies the p-periodic boundary condition. Moreover, $\tilde{u}(x,y) = \exp\{i(p-p_0)x\}\, u(x,y)$ in $\Omega_{0,h}$ and hence $\tilde{u}$ satisfies the generalized Dirichlet condition (i.e., $\tilde{u} \in L_2^{D,p,\ell oc}(\Omega)$) because $u \in L_2^{D,p_0,\ell oc}(\Omega)$. The preceding shows that $\tilde{u} \in D(A_p^{D,\ell oc})$. Finally the expansion (5.162) has the form (4.28) corresponding to $(p,\zeta) \in M$ and hence $\tilde{u} \in F_{p,\zeta,r}$.

Lemma 5.17. $J(p,\zeta,p_0,\zeta_0)$ maps $D(A_{p_0,\zeta_0,r})$ one-to-one onto $D(A_{p,\zeta,r})$.

Proof of Lemma 5.17. Lemma 5.16 implies that $J(p,\zeta,p_0,\zeta_0)$ maps $D(A_{p_0,\zeta_0,r})$ into $D(A_{p,\zeta,r})$. Moreover, it is clear from (5.149) and (5.141) that $J(p,\zeta,p_0,\zeta_0)$ is injective. The surjectivity may be verified by constructing the inverse. To do this let $v = P_{p,\zeta,r}u \in D(A_{p,\zeta,r})$ and

$$v(x,y) = \sum_{m\in Z} v_m(y) \exp\{i(p+m)x\} \text{ in } \Omega_{h,r} \tag{5.163}$$

and define

$$v_0(x,y) = \exp\{i(p_0-p)x\} \begin{cases} \sum_{m\in Z} d_m(p,\zeta,p_0,\zeta_0,y)^{-1}\, v_m(y) \exp\{i(p+m)x\} & \text{in } \Omega_{h,r}\ , \\ v(x,y) & \text{in } \Omega_{0,h}\ . \end{cases} \tag{5.164}$$

Note that $|d_m(p,\zeta,p_0,\zeta_0,y)^{-1}| < 2$ for all $(p,\zeta) \in N(p_0,\zeta_0)$, $m \in Z$ and $y \in R$. Hence the technique used to prove Lemma 5.16 can be used to show that $v_0 \in D(A_{p_0,\zeta_0,r})$ and $J(p,\zeta,p_0,\zeta_0)v_0 = v$.

Property (5.134) is obvious from definition (5.149) because $d_m(p_0,\zeta_0,p_0,\zeta_0,y) \equiv 1$. Hence the verification of properties (5.132)-(5.135) of J may be completed by proving

Lemma 5.18. $(p,\zeta) \to J(p,\zeta,p_0,\zeta_0) \in B(\mathcal{H},L_2^1(\Delta,\Omega_{0,r}))$ is continuous at (p_0,ζ_0).

Proof of Lemma 5.18. It must be shown that $\|J(p,\zeta,p_0,\zeta_0) - E\| \to 0$ when $(p,\zeta) \to (p_0,\zeta_0)$. An equivalent condition is

$$(5.165) \qquad \|J(p,\zeta,p_0,\zeta_0)v - Ev\|_{1,\Delta;0,r} \to 0 \text{ when } (p,\zeta) \to (p_0,\zeta_0) ,$$

uniformly for all $v \in \mathcal{H}$ such that $\|v\|_{1,\Delta;0,r} \le 1$ [13, p. 150]. To verify (5.165) define a bounded operator T_{p-p_0} in $L_2^1(\Delta,\Omega_{0,r})$ by

$$(5.166) \qquad T_{p-p_0} v(x,y) = \exp\{i(p-p_0)x\} v(x,y) .$$

Then for all $v \in \mathcal{H}$ one has

$$(5.167) \qquad \begin{aligned} \|J(p,\zeta,p_0,\zeta_0)v - Ev\|_{1,\Delta;0,r} &\le \|J(p,\zeta,p_0,\zeta_0)v - T_{p-p_0}v\|_{1,\Delta;0,r} \\ &\quad + \|T_{p-p_0}v - Ev\|_{1,\Delta;0,r} . \end{aligned}$$

Moreover one has, by (5.149) and (5.166),

$$(5.168) \qquad \begin{aligned} &J(p,\zeta,p_0,\zeta_0)\, v(x,y) - T_{p-p_0}v(x,y) \\ &= \begin{cases} \sum_{m\in Z} \{d_m(p,\zeta,p_0,\zeta_0,y) - 1\}\, v_m(y) \exp\{i(p+m)x\} & \text{in } \Omega_{h,r} , \\ 0 & \text{in } \Omega_{0,h} , \end{cases} \end{aligned}$$

whence

$$(5.169) \qquad \|J(p,\zeta,p_0,\zeta_0)v - T_{p-p_0}v\|_{1,\Delta;0,r} = 2\pi \sum_{m\in Z} \int_h^r I_m^1(y)\, dy$$

where

$$I_m^1(y) = (1 + |p+m|^2)\ |f_m\ v_m|^2 + |(D_y\ f_m)v_m + f_m\ D_y\ v_m|^2$$
(5.170)
$$+\ |f_m\ D_y^2\ v_m + 2\ D_y\ f_m\ D_y\ v_m + (D_y^2\ f_m)v_m - (p+m)^2\ f_m\ v_m|^2$$

and

(5.171) $$f_m = f_m(p,\zeta,p_0,\zeta_0,y) = d_m(p,\zeta,p_0,\zeta_0,y) - 1\ .$$

Now using the equicontinuity of the family $\{w_{p+m}(\zeta)\}$ and Lemma 5.14 it is not difficult to show that for each $\varepsilon > 0$ there is a neighborhood $N'(p_0,\zeta_0)$ of (p_0,ζ_0) in M such that

(5.172) $$0 \le I_m^1(y) \le \varepsilon^2\ I_m^0(y)$$

for all $(p,\zeta) \in N'(p_0,\zeta_0)$, $m \in Z$ and $h \le y \le r$, where $I_m^0(y)$ is defined by (5.153). It follows that (see (5.152))

(5.173) $$\|J(p,\zeta,p_0,\zeta_0)v - T_{p-p_0}v\|_{1,\Delta;0,r} \le \varepsilon\ \|v\|_{1,\Delta;0,r} \le \varepsilon$$

for all $v \in \mathcal{H}$ such that $\|v\|_{1,\Delta;0,r} \le 1$.

Similarly, an elementary calculation gives

(5.174) $$\|T_{p-p_0}v - Ev\|_{1,\Delta;0,r} \le \varepsilon$$

for all $v \in \mathcal{H}$ such that $\|v\|_{1,\Delta;0,r} \le 1$. Combining (5.167), (5.173) and (5.174) gives (5.165).

Construction of J - The Neumann Case. The mapping J defined by (5.149) is not applicable to the Neumann case because the operation $v \to \exp\{i(p-p_0)x\}v$ does not preserve the Neumann boundary condition. It will be shown that for grating domains $G \in S$ a suitable mapping J can be defined by replacing the multiplier $\exp\{i(p-p_0)x\}$ by a function of the form $\exp\{i(p-p_0)\ \phi(x,y)\}$. To this end note that if x_0 has property (1.9) of the definition of the class S then so do the points $x_0 + 2\pi m$, $m \in Z$. Moreover, it can be assumed that $x_0 = -\pi$ since equivalent domains are obtained by translating G parallel to the x-axis. This assumption is made in the remainder of this section. Also, to simplify the notation it will be assumed that

$$\partial G \cap \{(-\pi,y) \mid y \in R\} = (-\pi,y_0) \tag{5.175}$$

is a single point. The general case defined by (1.9) can be treated by the same method.

Property S implies that near $(-\pi,y_0)$ the boundary Γ has a representation $(x,y) = (f_1(s),f_2(s))$, where s is the arc length on Γ measured from $(-\pi,y_0)$, and $f_j \in C^3$. The vectors $\vec{t} = (f_1'(s),f_2'(s))$ and $\vec{n} = (-f_2'(s),f_1'(s))$ are unit tangent and normal vectors to Γ, respectively. The mapping $(s,t) \to (x,y)$ defined by

$$\begin{aligned} x &= f_1(s) - t\, f_2'(s) , \\ y &= f_2(s) + t\, f_1'(s) , \end{aligned} \tag{5.176}$$

has Jacobian 1 at $(s,t) = (0,0)$. Hence the inverse mapping

$$\begin{aligned} s &= \sigma(x,y) , \\ t &= \tau(x,y) , \end{aligned} \tag{5.177}$$

exists in a neighborhood of $(-\pi,y_0)$ and defines there a coordinate system of class C^2. The system is valid in a domain $\mathcal{O} = \{(s,t) : |s| < \delta_1, |t| < \delta_2\}$. It will be assumed that δ_1, δ_2 are chosen so small that $\mathcal{O} \subset \{(x,y) : |x + \pi| < \pi\}$. If extensions of $\sigma(x,y)$, $\tau(x,y)$ to $\mathcal{O} + (2\pi m,0)$ are defined by $\sigma(x + 2\pi m,y) = \sigma(x,y)$ and $\tau(x + 2\pi,y) = \tau(x,y)$ then the extended functions define coordinate systems in $\mathcal{O} + (2\pi m,0)$.

Introduce functions $\xi_j \in C(R)$ $(j = 1,2)$ such that $\xi_j(-\alpha) = \xi_j(\alpha)$, $\xi_j'(\alpha) \le 0$ and $\alpha \ge 0$ and

$$\xi_j(\alpha) = \begin{cases} 1 , & |\alpha| \le \delta_j/3 , \\ 0 , & |\alpha| \ge 2\delta_j/3 , \end{cases} \tag{5.178}$$

(whence $0 \le \xi_j(\alpha) \le 1$). The composite functions $\xi_1(\sigma(x,y))$ and $\xi_2(\tau(x,y))$ are then in class C^2. Similarly, introduce a function $\xi_3(x)$ such that

$$\xi_3(x) = \begin{cases} 1 , & |x + \pi| \le \delta_3/3 , \\ 0 , & 2\delta_3/3 \le |x + \pi| \le \delta_3 , \end{cases} \tag{5.179}$$

and

$$\xi_3(x + 2\pi) = \xi_3(x) \tag{5.180}$$

where $\delta_3 < \pi$. Finally define

$$\begin{aligned}\phi(x,y) &= (\sigma-\pi)\,\xi_1(\sigma)\,\xi_2(\tau) + x\,\xi_3(x)[1 - \xi_2(\tau)]\ , \quad -\pi \le x \le 0\ ,\\ \phi(x,y) &= (\sigma+\pi)\,\xi_2(\sigma)\,\xi_2(\tau) + x\,\xi_3(x)[1 - \xi_2(\tau)]\ , \quad 0 \le x \le \pi\ .\end{aligned} \tag{5.181}$$

The two parts of the definition are consistent because both give zero in a neighborhood of the y-axis. It will also be assumed that δ_2 is so small that $\xi_1(\sigma(\pm\pi,y)) = 1$ on the support of $\xi_2(\tau(\pm\pi,y))$.

The mapping J defined by (5.149) with $\exp\{i(p-p_0)x\}$ replaced by $\exp\{i(p-p_0)\,\phi(x,y)\}$ has the required properties (5.132)-(5.135). The proofs are the same as in the Dirichlet case except for the verification that $v' = J(p,\zeta,p_0,\zeta_0)v$ satisfies the Neumann and p-periodic boundary conditions. To verify the Neumann condition note that on the portion of Γ in the neighborhood defined by $\operatorname{supp}\phi \cap \{(x,y) : |\tau(x,y)| \le \delta_2/3\}$ one has

$$\phi(x,y) = (\sigma(x,y) \pm \pi)\,\xi_1(\sigma(x,y))\ . \tag{5.182}$$

Moreover, on the regular portion of Γ a simple calculation based on (5.176) gives

$$\sigma_x = \tau_y = f_1'(\sigma)\ , \quad \sigma_y = -\tau_x = f_2'(\sigma) \tag{5.183}$$

whence

$$D_\nu\,\sigma = (-f_2'(\sigma))\sigma_x + (f_1'(\sigma))\sigma_y = 0\ . \tag{5.184}$$

It follows from (5.182) and (5.184) that $v'(x,y)$ ($= \exp\{i(p-p_0)\phi(x,y)\}v(x,y)$ on Ω_h) satisfies

$$D_\nu\,v' = \exp\{i(p-p_0)\phi\}\,(D_\nu\,v + i(p-p_0)D_\nu\,\phi) = 0 \tag{5.185}$$

on $\operatorname{supp}\phi \cap \Gamma$. On the remainder of Γ $v' = v$ satisfies the generalized Neumann condition. The validity of the generalized Neumann condition for v' follows by a partition of unity argument.

To verify that v' satisfies the p-periodic boundary condition note that (5.181) and the assumption that $\xi_1(\sigma(\pm\pi,y)) = 1$ on the support of $\xi_2(\tau(\pm\pi,y))$ imply

$$\begin{aligned}\phi(\pi,y) &= (\sigma(\pi,y) + \pi)\ \xi_2(\tau(\pi,y)) + \pi(1 - \xi_2(\tau(\pi,y)))\\ &= (\sigma(-\pi,y) + \pi)\ \xi_2(\tau(-\pi,y)) + \pi(1 - \xi_2(\tau(-\pi,y)))\\ &= \phi(-\pi,y) + 2\pi\ \xi_2(\tau(-\pi,y)) + 2\pi(1 - \xi_2(\tau(-\pi,y)))\\ &= \phi(-\pi,y) + 2\pi\end{aligned} \tag{5.186}$$

and similarly

$$D_x\ \phi(\pi,y) = D_x\ \phi(-\pi,y)\ . \tag{5.187}$$

Thus

$$\begin{aligned}v'(\pi,y) &= \exp\ \{i(p-p_0)\ \phi(\pi,y)\}\ v(\pi,y)\\ &= \exp\ \{i(p-p_0)\ \phi(-\pi,y) + i(p-p_0)\ 2\pi + 2\pi i p_0\}\ v(-\pi,y)\\ &= \exp\ \{2\pi i p\}\ v'(-\pi,y)\end{aligned} \tag{5.188}$$

and similarly

$$D_x\ v'(x,y) = \exp\ \{i(p-p_0)\phi\}\ (D_x v + i(p-p_0)(D_x\phi)v) \tag{5.189}$$

whence

$$\begin{aligned}D_x v'(\pi,y) &= \exp\ \{i(p-p_0)\ \phi(-\pi,y) + i(p-p_0)2\pi + 2\pi i p_0\} \times\\ &\quad \times \{D_x\ v(-\pi,y) + i(p-p_0)\ D_x\ \phi(-\pi,y)\ v(-\pi,y)\}\\ &= \exp\ \{2\pi i p\}\ D_x\ v'(-\pi,y)\ .\end{aligned} \tag{5.190}$$

The above discussion completes the proof of the continuity of the family $\{A_{p,\zeta,r} : (p,\zeta) \in M\}$. The final assertion of Theorem 4.6 states that for fixed $p \in (-1/2,1/2]$ the family $\{A_{p,\zeta,r} : \zeta \in M_p\}$ is holomorphic in the generalized sense of Kato [13, p. 366]. This may be proved by means of the family of operators $J_p(\zeta,\zeta_0) \equiv J(p,\zeta,p,\zeta_0)$. It is only necessary to verify that $\zeta \to J_p(\zeta,\zeta_0)$ is holomorphic on M_p. A proof has been given by Alber [3, p. 271].

Proof of Theorem 4.7. $D(A_{p,\zeta,r}) = P_{p,\zeta,r}\, F_{p,\zeta,r}$ is a closed subspace of the Hilbert space $L_2^1(\Delta,\Omega_{0,r})$, by Theorem 4.4. $A_{p,\zeta,r} - z$ defines a bounded operator from this space into $L_2(\Omega_{0,r})$. Thus the operator $T : L_2(\Omega_{0,r}) \to D(A_{p,\zeta,r})$ defined by $Tf = R(A_{p,\zeta,r},z)f$ for all $f \in L_2(\Omega_{0,r})$ is closed and defined on all of $L_2(\Omega_{0,r})$. Thus T is bounded, by the closed graph theorem.

Next note that $R(A_{p,\zeta,r},z) = E\,T$ where $E : D(A_{p,\zeta,r}) \to L_2(\Omega_{0,r})$ is the natural embedding. Hence, the compactness of the resolvent of $A_{p,\zeta,r}$ follows from the compactness of E. Now, in the Neumann case $F_{p,\zeta,r} \subset L_2^{1,p,\ell oc}(\Omega)$ and hence the compactness of E follows from the hypothesis $G \in LC$. In the Dirichlet case, $F_{p,\zeta,r} \subset L_2^{D,p,\ell oc}(\Omega)$ = closure of $C_p^\infty(\Omega)$ in $L_2^{1,\ell oc}(\Omega)$. The last set can be regarded as a subset of $L_2^{1,p,\ell oc}(B_0)$ for which the natural embedding into $L_2^{\ell oc}(B_0)$ has the local compactness property. Hence, in this case E is compact without local restrictions on $\Gamma = \partial G \cap \Omega$. This proves the compactness of the resolvent of $A_{p,\zeta,r}$. The discreteness of $\sigma(A_{p,\zeta,r})$ follows immediately; see Kato [13, p. 187].

Proof of Theorem 4.8. It will be shown that if $\zeta \in M_p^+$ then the operator in $L_2(\Omega_{0,r})$ defined by

$$T = P_{p,\zeta,r}\, R(A_p,\pi_p(\zeta))\, P_r \tag{5.191}$$

is a bounded inverse of $A_{p,\zeta,r} - \pi_p(\zeta)$ in $L_2(\Omega_{0,r})$. To prove that T is a right inverse of $A_{p,\zeta,r} - \pi_p(\zeta)$ let $f \in L_2(\Omega_{0,r})$ and define $u = R(A_p,\pi_p(\zeta))f$. Then $u \in R(A_p)$ and

$$(A_p - \pi_p(\zeta))u = P_r f = \begin{cases} f \text{ in } \Omega_{0,r}, \\ 0 \text{ in } \Omega_r . \end{cases} \tag{5.192}$$

In particular, $(\Delta + \pi_p(\zeta))u = 0$ in Ω_r and thus since $u \in L_2(\Omega)$ the Fourier expansion (4.28) must hold with $\operatorname{Im} w_{p+m}(\zeta) > 0$ for all $m \in Z$. Thus $u \in F_{p,\zeta,r}$ and it follows that $P_{p,\zeta,r}u \in D(A_{p,\zeta,r})$ and

$$[A_{p,\zeta,r} - \pi_p(\zeta)]\,T\,f = [A_{p,\zeta,r} - \pi_p(\zeta)]P_{p,\zeta,r}u = (-\Delta - \pi_p(\zeta))u\Big|_{\Omega_{0,r}} = f \tag{5.193}$$

by (5.192).

To prove that T is a left inverse of $A_{p,\zeta,r} - \pi_p(\zeta)$ let $v \in D(A_{p,\zeta,r})$. Then $u = P^{-1}_{p,\zeta,r} v$ has Fourier expansion (4.28) with $\operatorname{Im} w_{p+m}(\zeta) > 0$ for all $m \in Z$ because $\zeta \in M_p^+$. Thus $u \in F_{p,\zeta,r} \cap L_2^1(\Delta,\Omega) = D(A_p)$ and one has

$$
\begin{aligned}
P_r[A_{p,\zeta,r} - \pi_p(\zeta)]v &= (-\Delta - \pi_p(\zeta))P^{-1}_{p,\zeta,r} v \\
&= (A_p - \pi_p(\zeta))P^{-1}_{p,\zeta,r} v
\end{aligned}
\tag{5.194}
$$

whence $T[A_{p,\zeta,r} - \pi_p(\zeta)]v = v$.

Proof of Theorem 4.9. The family of operators $\{A_{p,\zeta,r} - \pi_p(\zeta) \mid \zeta \in M_p\}$ is holomorphic (Theorem 4.6) and has compact resolvents (Theorem 4.7). It follows from a theorem of Kato [13, p. 371] that either $\Sigma_p = M_p$ or Σ_p has no accumulation points in M_p. But $M_p^+ \cap \Sigma_p = \phi$ by Theorem 4.8. Hence the second alternative must hold.

To prove that Σ_p is independent of $r > h$ let $h < r' < r$ and suppose that $\pi_p(\zeta) \in \sigma(A_{p,\zeta,r})$. Then there exists a non-zero $v \in D(A_{p,\zeta,r})$ such that $(A_{p,\zeta,r} - \pi_p(\zeta))v = 0$ in $L_2(\Omega_{0,r})$. But then $u = P^{-1}_{p,\zeta,r} u \in F_{p,\zeta,r} \subset L_2^{1,\ell oc}(\Delta,\Omega)$ and $(\Delta + \pi_p(\zeta))u = 0$ in all of Ω. In particular, the Fourier expansion (4.28) holds in $\Omega_{r',\infty}$. Thus $u \in F_{p,\zeta,r'}$ and hence $P_{p,\zeta,r'} u \in D(A_{p,\zeta,r'})$ and $(A_{p,\zeta,r'} - \pi_p(\zeta))P_{p,\zeta,r'} u = 0$. Thus $\pi_p(\zeta) \in \sigma(A_{p,\zeta,r'})$ as was to be shown. The same argument is applicable if $r' > r$.

Proof of Corollary 4.10. Theorem 4.7 implies that every $z \in C$ is either an eigenvalue of $A_{p,\zeta,r}$ or lies in $\rho(A_{p,\zeta,r})$. Hence for each $\zeta \in M_p - \Sigma_p$ one has $\pi_p(\zeta) \in \rho(A_{p,\zeta,r})$ and it follows from [13, p. 367] that $R_{p,\zeta,r}$ is holomorphic on $M_p - \Sigma_p$. Thus to complete the proof it is enough to show that each $\zeta_0 \in \Sigma_p$ is a pole of $R_{p,\zeta,r}$. This will be deduced from S. Steinberg's theorem [24] and the following

Lemma 5.19. Let $\zeta \in M_p^+$ and $\operatorname{Im} \pi_p(\zeta) > 0$ (resp., < 0). Then every $z \in \sigma(A_{p,\zeta,r})$ satisfies $\operatorname{Im} z < 0$ (resp. > 0).

Proof of Lemma 5.19. Let $v \in D(A_{p,\zeta,r})$ be an eigenfunction of $A_{p,\zeta,r}$ with eigenvalue z : $v \neq 0$ and $A_{p,\zeta,r} v = zv$. Then $u = P^{-1}_{p,\zeta,r} v \in F_{p,\zeta,r}$ and hence

$$
(\Delta + z)u = (\Delta + z)v = 0 \text{ in } \Omega_{0,r}, \text{ and} \tag{5.195}
$$

$$
(\Delta + z')u = 0 \text{ in } \Omega_{r,\infty} \tag{5.196}
$$

where $z' = \pi_p(\zeta)$. Moreover, Lemma 4.1 and Sobolev's embedding theorems imply that $y \to u(\cdot,y)$ is in $C^1([h,\infty),L_2(-\pi,\pi))$. In addition, the assumption $\zeta \in M_p^+$ implies that $u \in D(A_p) \subset L_2(\Omega)$.

Application of Green's theorem to u and $\bar{u}$ in $\Omega_{r,\infty}$ gives, by (5.196),

$$(5.197)\qquad (-2i \text{ Im } z') \int_{\Omega_{r,\infty}} |u|^2 \, dX = -\int_{-\pi}^{\pi} \left\{\bar{u} \frac{u}{y} - u \frac{\bar{u}}{y}\right\}_{y=r} dx \ .$$

Similarly, application of Green's theorem in $\Omega_{0,r}$ gives

$$(5.198)\qquad (-2i \text{ Im } z) \int_{\Omega_{0,r}} |u|^2 \, dX = \int_{-\pi}^{\pi} \left\{\bar{u} \frac{u}{y} - u \frac{\bar{u}}{y}\right\}_{y=r} dx \ .$$

Adding (5.197) and (5.198) gives

$$(5.199)\qquad \text{Im } z' \int_{\Omega_{r,\infty}} |u|^2 \, dX + \text{Im } z \int_{\Omega_{0,r}} |u|^2 \, dX = 0 \ .$$

Thus if $\text{Im } \pi_p(\zeta) = \text{Im } z' > 0$ and $\text{Im } z \geq 0$ then $u(X) \equiv 0$ in $\Omega_{r,\infty}$. But then $u(\cdot,r) = 0$ and $D_y\, u(\cdot,r) = 0$ and hence $u(X) \equiv 0$ in $\Omega_{0,r}$ by the unique continuation property for (5.195). Hence $\text{Im } \pi_p(\zeta) > 0$ implies $\text{Im } z < 0$. The other case is proved in the same way.

Returning to the proof of Corollary 4.10, it will be shown first that every $\zeta_0 \in \Sigma_p$ such that

$$(5.200)\qquad \text{Im } \pi_p(\zeta_0) \geq 0$$

is a pole of $R_{p,\zeta,r}$. To this end choose $\zeta_1 \in M_p^+$ such that $\text{Im } \pi_p(\zeta_1) > 0$, so that

$$(5.201)\qquad \{z \mid \text{Im } z \geq 0\} \subset \rho(A_{p,\zeta_1,r})$$

by Lemma 5.19. Next choose a $z_1 \in C$ such that

$$(5.202)\qquad z_1 \in \rho(A_{p,\zeta_1,r}) \ ,$$

$$(5.203)\qquad z_1 \in \rho(A_{p,\zeta,r}) \text{ for all } \zeta \in N(\zeta_0,\delta) \ ,$$

where $N(\zeta_0,\delta)$ is the component of $\pi_p^{-1}(D(\pi_p(\zeta_0),\delta))$ containing ζ_0. $N(\zeta_0,\delta)$ has compact closure and hence such numbers z_1 exist by Corollary 5.5 above. In the remainder of the proof the following notation is used:

$$(5.204)\qquad \begin{cases} R(\zeta,z) = (A_{p,\zeta,r} - z)^{-1} , \\ R(\zeta) = R(\zeta,\pi_p(\zeta)) . \end{cases}$$

With the above choices of ζ_1 and z_1 the operator

$$(5.205)\qquad B(z) = (1 - (z - z_1)\, R(\zeta_1,z_1))^{-1}$$

exists and is holomorphic for Im $z \geq 0$ (i.e., in an open set containing Im $z \geq 0$). Indeed,

$$(5.206)\qquad 1 - (z - z_1)\, R(\zeta_1,z_1) = (A_{p,\zeta_1,r} - z)\, R(\zeta_1,z_1)$$

and the existence of $B(z)$ follows from (5.201). The analyticity follows from that of $R(\zeta_1,z)$.

To complete the proof of Corollary 4.10 note that (5.200) and (5.201) imply that $\pi_p(\zeta_0) \in \rho(A_{p,\zeta_1,r})$. Since the resolvent set is open, the continuity of π_p implies that there exists a $\delta > 0$ such that $\pi_p(\zeta) \in \rho(A_{p,\zeta_1,r})$ for all $\zeta \in N(\zeta_0,\delta)$. Hence $B(\pi_p(\zeta))$ exists and is holomorphic in $N(\zeta_0,\delta)$. Now for all such ζ one has, by (5.202), (5.203),

$$(5.207)\qquad \begin{aligned} 1 - (\pi_p(\zeta) - z_1)\, R(\zeta,z_1) = {} & 1 - (\pi_p(\zeta) - z_1)\, R(\zeta_1,z_1) \\ & - (\pi_p(\zeta) - z_1)\, \{R(\zeta,z_1) - R(\zeta_1,z_1)\} . \end{aligned}$$

Multiplying by $B(\pi_p(\zeta))$ gives

$$(5.208)\qquad B(\pi_p(\zeta))\, \{1 - (\pi_p(\zeta) - z_1)\, R(\zeta,z_1)\} = 1 - T(\zeta)$$

where

$$(5.209)\qquad T(\zeta) = (\pi_p(\zeta) - z_1)\, B(\pi_p(\zeta))\, \{R(\zeta,z_1) - R(\zeta_1,z_1)\}$$

defines a compact operator-valued holomorphic family in $N(\zeta_0,\delta)$. By Steinberg's theorem [24], $(1 - T(\zeta))^{-1}$ either exists nowhere or is meromorphic in $N(\zeta_0,\delta)$. The second case must hold because the singularities of $(1 - T(\zeta))^{-1}$ are those of $R(\zeta)$ and hence are isolated. In particular, for δ small enough $(1 - T(\zeta))^{-1}$ is analytic in $N(\zeta_0,\delta)$ except for a pole at $\zeta = \zeta_0$. Equation (5.208) then implies

(5.210) $$A_{p,\zeta,r} - \pi_p(\zeta) = B(\pi_p(\zeta))^{-1} (1 - T(\zeta))\, R(\zeta,z_1)^{-1}$$

and therefore

(5.211) $$R(\zeta) = R(\zeta,z_1)(1 - T(\zeta))^{-1}\, B(\pi_p(\zeta))$$

for $\zeta \in N(\zeta_0,\delta) - \{\zeta_0\}$. This exhibits $R(\zeta)$ as a product of operators that are holomorphic at ζ_0 and one that has a pole there. The residue of $R(\zeta)$ at ζ_0 has finite rank [24] and hence $\pi_p(\zeta_0)$ is an eigenvalue of finite algebraic multiplicity [13, p. 181].

Proof of Corollary 4.11. This result follows immediately from Theorems 4.4 and 4.8.

Proof of Corollary 4.12. It will be shown that

(5.212) $$\sigma_0(A_p) \subset \pi_p(\overline{M_p^+} \cap \Sigma_p) .$$

The discreteness of $\sigma_0(A_p)$ will then follow from Theorem 4.9. To prove (5.212) let $\lambda \in \sigma_0(A_p) \subset \sigma(A_p) = [p^2,\infty)$ and let $\lambda \pm i0$ denote the points of $\overline{M_p^+}$ above λ so

(5.213) $$\pi_p(\lambda \pm i0) = \lambda .$$

If $u \in D(A_p)$ is a corresponding eigenfunction of A_p then $u \in F_{p,\lambda\pm i0,r}$, $v_\pm = P_{p,\lambda\pm i0,r}u \in D(A_{p,\lambda\pm i0,r})$ and $(A_{p,\lambda\pm i0,r} - \lambda)v_\pm = 0$. Thus $(A_{p,\lambda\pm i0,r} - \pi_p(\lambda \pm i0))$ is not invertible and hence $\lambda \pm i0 \in \Sigma_p$.

The inclusion (5.212) and Theorem 4.9 imply that $\sigma_0(A_p)$ has no finite limit points. To show that each $\lambda \in \sigma_0(A_p)$ has a finite dimensional eigenspace note that the algebraic and geometric eigenspaces of A_p coincide because A_p is selfadjoint. Moreover $P_{p,\lambda\pm i0,r}$ maps the eigenspace of $\lambda \in \sigma_0(A_p)$ onto the geometric eigenspace of $A_{p,\lambda\pm i0,r}$ for λ, as was shown above. However, the latter coincides with the geometric eigenspace of the compact operator $R(\lambda \pm i0,z)$ defined by (5.204) and hence is finite dimensional.

Proof of Corollary 4.13. To prove (4.43) note that if $\lambda \in \pi_p(\overline{M_p^+} \cap \Sigma_p) - T_p$ then $\lambda + i0$ or $\lambda - i0$ is in $\overline{M_p^+} \cap \Sigma_p$ and hence $\lambda = \pi_p(\lambda \pm i0)$ is an eigenvalue of $A_{p,\lambda+i0,r}$ or $A_{p,\lambda-i0,r}$ with eigenfunction v_+ or v_-. But then $u_+ = P^{-1}_{p,\lambda+i0,r}\, v_+$ or $u_- = P^{-1}_{p,\lambda-i0,r}\, v_-$ will have a p-periodic extension to G that is a pure outgoing or incoming R-B wave for A. It follows from Theorem 2.1 that u_+ or u_- is an eigenfunction for A_p with eigenvalue λ; i.e., $\lambda \in \sigma_0(A_p)$.

Proof of Theorem 4.14. Both statements of Theorem 4.14 follow from the continuity of the family $\{A_{p,\zeta,r} : (p,\zeta) \in M\}$ and a theorem of Kato [17, Theorem IV.2.25]. Indeed, if $(p_0,\zeta_0) \in M - \Sigma$ then $\pi_{p_0}(\zeta_0) \in \rho(A_{p_0,\zeta_0,r})$ and hence $R_{p,\zeta,r} \to R_{p_0,\zeta_0,r}$ when $(p,\zeta) \to (p_0,\zeta_0)$. Moreover, it follows from Kato's theorem that there exists a neighborhood $N(p_0,\zeta_0,\rho,\delta) \subset M - \Sigma$.

Proof of Theorem 4.15. This result is an immediate corollary of Theorem 4.14 and Theorem 4.4.

Proof of Corollary 4.16. Theorem 4.15 implies that $(p,\zeta) \to P^{-1}_{p,\zeta,r} R_{p,\zeta,r} \in B(L_2(\Omega_{0,r}), L_2^1(\Delta,\Omega_{0,r'}))$ is continuous on $M - \Sigma$ for each $r' > r$. This implies (4.49) with

$$C(K,r,r') = \operatorname*{Max}_{(p,\zeta)\in K} \|P^{-1}_{p,\zeta,r} R_{p,\zeta,r}\|_{r,r'} \tag{5.214}$$

where $\|\cdot\|_{r,r'}$ denotes the operator norm in the space $B(L_2(\Omega_{0,r}), L_2^1(\Delta,\Omega_{0,r'}))$.

Proof of Corollary 4.17. This result is a special case of Corollary 4.16.

§6. The Eigenfunction Expansions for A_p

This section presents a construction, based on the limiting absorption theorem of §4, of the diffracted plane wave eigenfunctions $\phi_\pm(X,p+m,q)$ and a derivation of the corresponding eigenfunction expansions for A_p. For brevity the derivation is restricted to the cases for which $\sigma_0(A_p) = \phi$. The modifications that are needed when $\sigma_0(A_p) \neq \phi$ are indicated at the end of the section.

Throughout this section $p \in (-1/2,1/2]$ is fixed, $m \in Z$ and $q > 0$. $\phi_{0\pm}(X,p+m,q)$ denotes the generalized eigenfunction for $A_{0,p}$; that is, one of the functions (3.25), (3.26). The corresponding outgoing and incoming diffracted plane waves for A_p are characterized by the properties

$$\phi_\pm(\cdot,p+m,q) \in D(A_p^{\ell oc}) , \tag{6.1}$$

$$(\Delta + \omega^2(p+m,q))\, \phi_\pm(X,p+m,q) = 0 \text{ in } \Omega , \tag{6.2}$$

$$(6.3)\qquad \phi_{\pm}(X,p+m,q) = \phi_{0\pm}(X,p+m,q) + \phi'_{\pm}(X,p+m,q)\ , \quad y \geq h\ ,$$

where ϕ'_+ (resp., ϕ'_-) is an outgoing (resp., incoming) diffracted plane wave in Ω_h. These properties imply the symmetry relation

$$(6.4)\qquad \phi_-(X,p+m,q) = \overline{\phi_+(X,-p-m,q)}\ .$$

Hence it will be sufficient to construct the functions $\phi_+(X,p+m,q)$.

To construct ϕ_+ let $r > h$ be fixed and introduce a function $j \in C^\infty[0,\infty)$ such that $j'(y) \geq 0$, $0 \leq j(y) \leq 1$, $j(y) \equiv 0$ for $0 \leq y \leq (h+r)/2$ and $j(y) \equiv 1$ for $y \geq r$. Next define the function $\phi'_+(X,p+m,q)$ for all $X \in \Omega$ by

$$(6.5)\qquad \phi_+(X,p+m,q) = j(y)\ \phi_0(X,p+m,q) + \phi'_+(X,p+m,q)\ , \quad X \in \Omega\ .$$

Then (6.1), (6.2), (6.3) imply that ϕ'_+ is characterized by the properties

$$(6.6)\qquad \phi'_+(\cdot,p+m,q) \in D(A_p^{\ell oc})$$

$$(6.7)\qquad (\Delta + \omega^2(p+m,q))\ \phi'_+(X,p+m,q) = -M(X,p+m,q) \text{ in } \Omega\ ,$$

$$(6.8)\qquad \phi'_+(X,p+m,q) \text{ is an outgoing diffracted plane wave}\ .$$

The function M in (6.7) is defined for all $X \in R_0^2$, $p + m \in R$ and $q > 0$ by

$$(6.9)\qquad \begin{aligned} M(X,p+m,q) &= (\Delta + \omega^2(p+m,q))\ j(y)\ \phi_0(X,p+m,q) \\ &= j''(y)\ \phi_0(X,p+m,q) + 2\ j'(y)\ D_2\ \phi_0(X,p+m,q) \end{aligned}$$

and has the properties

$$(6.10)\qquad M \in C^\infty(R_0^2 \times R \times R_0)\ ,$$

$$(6.11)\qquad M(x+2\pi,y,p+m,q) = \exp\{2\pi i p\}\ M(x,y,p+m,q)\ ,$$

$$(6.12)\qquad \operatorname{supp} M(\cdot,p+m,q) \subset \{X \mid (h+r)/2 \leq y \leq r\}\ .$$

It follows that $M(\cdot,p+m,q)|_{\Omega_{0,r}} \in L_2(\Omega_{0,r})$ and hence (6.6), (6.7), (6.8) can be integrated by means of the analytic continuation of the resolvent of A_p defined by (4.50). More generally

$$(6.13)\qquad \phi'(\cdot,p+m,q,z) = P^{-1}_{p,z,r}\, R_{p,z,r}\, M(\cdot,p+m,q) \in D(A_p^{\ell oc})$$

and $z \to \phi'(\cdot,p+m,q,z) \in L_2^{1,\ell oc}(\Delta,\Omega)$ is continuous for all $q > 0$ and $z \in \overline{M_p^{\mp}} - \Sigma_p$. Hence, $\phi'(\cdot,p+m,p,\lambda+i0) \in D(A_p^{\ell oc})$ satisfies $(\Delta + \lambda)\phi' = -M$ in Ω and the outgoing radiation condition (4.56) for all $\lambda \in [p^2,\infty) - T_p$. In particular, the solution of (6.6), (6.7), (6.8) is defined by

$$(6.14)\qquad \phi_+'(\cdot,p+m,q) = \phi'(\cdot,p+m,q,\omega^2(p+m,q) + i0)$$

for all $q \in R_0 - E_{m,p}$ where

$$(6.15)\qquad E_{m,p} = \{q > 0 \mid \omega^2(p+m,q) \in T_p\}\ .$$

Note that $E_{m,p}$ is a countable subset of $R_0 = (0,\infty)$ with no finite limit points.

The diffracted plane wave $\phi_+(X,p+m,q)$ is defined by (6.5), (6.13) and (6.14) and one has

<u>Theorem 6.1</u>. Let G be a grating domain of the class defined in §1 and let $\sigma_0(A_p) = \phi$. Then there exist unique diffracted plane wave eigenfunctions $\phi_\pm(X,p+m,q)$ for each $p \in (-1/2,1/2]$, $m \in Z$ and $q \in R_0 - E_{m,p}$. Moreover, $q \to \phi_\pm(\cdot,p+m,q) \in L_2^{1,\ell oc}(\Delta,\Omega)$ is continuous for $q \in R_0 - E_{m,p}$.

The uniqueness follows from Theorem 2.1 and $\sigma_0(A_p) = \phi$. The continuity is a consequence of Theorem 4.15.

The functions

$$(6.16)\qquad \phi(X,p+m,q,z) = j(y)\,\phi_0(X,p+m,q) + \phi'(X,p+m,q,z) \in D(A_p^{\ell oc})\ ,$$

which are defined for $p \in (-1/2,1/2]$, $m \in Z$, $q > 0$ and $z \in \overline{M_p^+} - \Sigma_p$ will be used in deriving the eigenfunction expansions for ϕ_+ and ϕ_-. They will be called approximate eigenfunctions of A_p because

$$(6.17)\qquad (\Delta + z)\,\phi(X,p+m,q,z) = (z - \omega^2(p+m,q))\, j(y)\,\phi_0(X,p+m,q)$$

and

$$(6.18)\qquad \phi(X,p+m,q,\omega^2(p+m,q) \pm i0) = \phi_\pm(X,p+m,q)\ .$$

Construction of the Spectral Family of A_p. The selfadjoint operator A_p in $L_2(\Omega)$ has a spectral family $\{\Pi_p(\mu) \mid \mu \geq p^2\}$ which is continuous when $\sigma_0(A_p) = \phi$. The spectral measure $\Pi_p(I) = \Pi_p(b) - \Pi_p(a)$ of an interval $I = [a,b]$ will now be calculated by means of Stone's formula

$$(6.19) \qquad \|\Pi_p(I)f\|^2 = \lim_{\sigma\to 0+} \frac{\sigma}{\pi}\int_I \|R(A_p,\lambda \pm i\sigma)f\|^2 \, d\lambda$$

and the eigenfunctions $\phi_\pm$. Only the main steps of the calculation will be given because a detailed presentation of the analogous calculation for exterior domains was given in [30].

To begin it will be assumed that $I \subset [p^2,\infty) - T_p$ and $f \in L_2^{com}(\Omega)$. Note that if $j(y)$ is the cut-off function of (6.5) then

$$(6.20) \qquad |(1 - j^2(y))\, R(A_p,z)\, f(X)|^2 \leq \chi_r(y)\, |R(A_p,z)\, f(X)|^2$$

where χ_r is the characteristic function of $[0,r]$. Since $\lim_{\sigma\to 0+} R(A_p,\lambda\pm i\sigma)f$ exists in $L_2(\Omega_{0,r})$, uniformly for $\lambda \in I$, it follows that

$$(6.21) \qquad \int_\Omega (1 - j^2(y))\, |R(A_p,\lambda\pm i\sigma)\, f(X)|^2 \, dX = O(1)\ , \quad \sigma \to 0+\ ,$$

uniformly for $\lambda \in I$. Define a linear operator $J : L_2(\Omega) \to L_2(B_0)$ by

$$(6.22) \qquad J\, f(X) = \begin{cases} j(y)\, f(X)\ , & X \in \Omega \\ 0\ , & X \in B_0 - \Omega\ . \end{cases}$$

Then $\|J\| = 1$ and (6.21) implies

$$(6.23) \qquad \|R(A_p,z)f\|^2 = \|J\, R(A_p,z)f\|^2 + O(1)\ , \quad \text{Im } z \to 0\ ,$$

uniformly for Re $z \in I$. Next, Parseval's relation (3.29) for $A_{0,p}$ and (6.23) imply

$$(6.24) \quad \|R(A_p,z)f\|^2 = \sum_{m\in Z} \|(J\, R(A_p,z)f)_0^\sim\,(p+m,\cdot)\|^2 + O(1),\ \text{Im } z \to 0\ ,$$

uniformly for Re $z \in I$. To relate this to the eigenfunctions $\phi_\pm$ define

$$(6.25) \qquad \tilde{f}(p+m,q,z) = \int_\Omega \overline{\phi(X,p+m,q,\bar{z})}\, f(X)\, dX\ ,\quad f \in L_2^{com}(\Omega)\ ,$$

and note

Lemma 6.2. For all $f \in L_2^{com}(\Omega)$ one has

$$\tilde{f}(p+m,q,z) = (\omega^2(p+m,q) - z)\ (J\ R(A_p,z)f)_0^{\sim}\ (p+m,q)\ . \tag{6.26}$$

A heuristic proof of (6.26) is contained in the following formal calculations, based on (6.17).

$$\begin{aligned} \tilde{f}(p+m,q,z) &= \int_\Omega \overline{R(A_p,\overline{z})(A_p - \overline{z})\ \phi(X,p+m,q,\overline{z})}\ f(X)\ dx \\ &= \int_\Omega \overline{(\omega^2(p+m,q) - z)\ j(y)\ \phi_0(X,p+m,q)}\ R(A_p,z)\ f(X)\ dx \\ &= (\omega^2(p+m,q) - z) \int_{B_0} \overline{\phi_0(X,p+m,q)}\ j(y)\ R(A_p,z)\ f(X)\ dx \\ &= (\omega^2(p+m,q) - z)\ (J\ R(A_p,z)f)_0^{\sim}\ (p+m,q)\ . \end{aligned} \tag{6.27}$$

The calculation is not rigorous because the presence of the term $j\phi_0$ in (5.16) implies that $\phi(\cdot,p+m,q,z) \notin D(A_p)$. A rigorous but longer proof may be given by the technique of [30, p. 94].

Combining (6.24) and (6.26) gives

$$\|R(A_p,z)f\|^2 = \sum_{m\in Z} \left\| \frac{\tilde{f}(p+m,\cdot,z)}{\omega^2(p+m,\cdot)-z} \right\|^2 + O(1)\ ,\quad \text{Im } z \to 0\ , \tag{6.28}$$

uniformly for Re $z \in I$. Hence, putting $z = \lambda \pm i\sigma$, multiplying by σ/π and integrating over $\lambda \in I$ gives

$$\begin{aligned} \frac{\sigma}{\pi} \int_I \|R(A_p,\lambda\pm i\sigma)f\|^2\ d\lambda &= \frac{\sigma}{\pi} \int_I \sum_{m\in Z} \int_0^\infty \frac{|\tilde{f}(p+m,q,\lambda\pm i\sigma)|^2}{(\lambda-\omega^2(p+m,q))^2+\sigma^2}\ dq\ d\lambda + O(\sigma) \\ &= \sum_{m\in Z} \int_0^\infty \left(\frac{\sigma}{\pi} \int_I \frac{|\tilde{f}(p+m,q,\lambda\pm i\sigma)|^2}{(\lambda-\omega^2(p+m,q))^2+\sigma^2}\ d\lambda \right) dq + O(\sigma) \end{aligned} \tag{6.29}$$

by Fubini's theorem. The determination of $\Pi_p(I)$ will be completed by calculating the limit for $\sigma \to 0$ of the last equation. Note that the continuity of the approximate eigenfunctions (6.16) for $q > 0$, $z \in \overline{M_p^{\mp}} - \Sigma_p$ (cf. (6.13)) implies that $\tilde{f}(p+m,q,\lambda\pm i\sigma)$ is continuous for $q > 0$, $\lambda \in [p^2,\infty) - T_p$, $\sigma \geq 0$. Thus if one defines

$$\tilde{f}_\pm(p+m,q) = \tilde{f}(p+m,q,\omega^2(p+m,q) \mp i0),\ q \in R_0 - E_{m,p} \tag{6.30}$$

then for all $f \in L_2^{com}(\Omega)$

$$\tilde{f}_{\pm}(p+m,q) = \int_{\Omega} \overline{\phi_{\pm}(X,p+m,q)}\, f(X)\, dx \tag{6.31}$$

and

$$\tilde{f}_{\pm}(p+m,\cdot) \in C(R_0 - E_{m,p}) . \tag{6.32}$$

The calculation of the limiting form of (6.29) will be based on the following two lemmas.

Lemma 6.3. For every $f \in L_2^{com}(\Omega)$ and every closed interval $I \subset [p^2,\infty) - T_p$ one has

$$\lim_{\sigma\to 0+} \frac{\sigma}{\pi} \int_I \frac{|\tilde{f}(p+m,q,\lambda\pm i\sigma)|^2}{(\lambda-\omega^2(p+m,q))^2+\sigma^2}\, d\lambda = \chi_I(\omega^2(p+m,q))\, |\tilde{f}_{\mp}(p+m,q)|^2 \tag{6.33}$$

for all $q \in R_0^2 - E_{m,p}$ where $\chi_I(\lambda)$ is the characteristic function of I, normalized so that $\chi_I(a) = \chi_I(b) = 1/2$.

Lemma 6.3 follows from the continuity of $\tilde{f}(p+m,q,\lambda\pm i\sigma)$ and well-known properties of the Poisson kernels; cf. [30, p. 101].

Lemma 6.4. For every $f \in L_2^{com}(\Omega)$, every $p \in (1/2,1/2]$, every closed interval $I \subset [p^2,\infty) - T_p$ and every $\sigma_0 > 0$ there exists a constant $C = C(f,p,I,\sigma_0)$ such that

$$\sum_{m\in Z} \int_0^{\infty} |\tilde{f}(p+m,q,\lambda\pm i\sigma)|^2\, dq \le C \tag{6.34}$$

for all $\lambda \in I$ and $\sigma \in [0,\sigma_0]$.

This result is the analogue of [30, Lemma 6.8, p. 103]. A full proof, based on Corollary 4.17, is given in §7 below.

The limit of equation (6.29) for $\sigma \to 0$ may now be calculated. Lemma 6.3 gives the limits of the inner integrals in (6.29). Term-wise passage to the limit can be justified by Lemma 6.4 and Lebesgue's dominated convergence theorem; see [30] for details. The result is, by (6.19),

$$\|\Pi_p(I)f\|^2 = \sum_{m\in Z} \int_0^{\infty} \chi_I(\omega^2(p+m,q))\, |\tilde{f}_{\pm}(p+m,q)|^2\, dq \tag{6.35}$$

for all $f \in L_2^{com}(\Omega)$ and $I \subset [p^2,\infty) - T_p$ where $\tilde{f}_{\pm}(p+m,q)$ is given by (6.31).

The Eigenfunction Expansions for A_p. The eigenfunction expansions for A_p based on $\phi_+(X,p+m,q)$ and $\phi_-(X,p+m,q)$ can be derived from (6.35) and the spectral theorem by standard methods; cf. [30, p. 109ff]. Only the results are given here. Details may be found in [30].

To begin note that since $\sigma_0(A_p) = \phi$ the restriction $I \subset [p^2,\infty) - T_p$ can be dropped; (6.35) is valid for $f \in L_2^{com}(\Omega)$ and all $I \subset [p^2,\infty)$. Making $I \to [p^2,\infty)$ then gives the Parseval relation

$$(6.36) \qquad \|f\|^2_{L_2(\Omega)} = \sum_{m\in Z} \|\tilde{f}_\pm(p+m,\cdot)\|^2_{L_2(R_0)}$$

for all $f \in L_2^{com}(\Omega)$. Together with (6.32) this implies that for $f \in L_2^{com}(\Omega)$,

$$(6.37) \qquad \tilde{f}_\pm(p+m,\cdot) \in C(R_0 - E_{m,p}) \cap L_2(R_0) \ .$$

A standard density argument then implies

<u>Theorem 6.5</u>. For all $f \in L_2(\Omega)$ the limits

$$(6.38) \qquad \tilde{f}_\pm(p+m,q) = \underset{M\to\infty}{\ell.i.m.} \int_{\Omega_{0,M}} \overline{\phi_\pm(X,p+m,q)}\, f(X)\, dX$$

exist in $L_2(R_0)$ and (6.35), (6.36) are valid for all $f \in L_2(\Omega)$.

An eigenfunction representation of the spectral family can now be obtained from (6.35) by the usual polarization and factorization arguments. In this way one obtains

<u>Theorem 6.6</u>. For all $f \in L_2(\Omega)$ one has

$$(6.39) \quad \Pi_p(\mu)\, f(X) = \sum_{(p+m)^2\le\mu} \int_0^{(\mu-(p+m)^2)^{1/2}} \phi_\pm(X,p+m,q)\, \tilde{f}_\pm(p+m,q)\, dq$$

and hence

$$(6.40) \qquad f(X) = \underset{M\to\infty}{\ell.i.m.} \sum_{|m|\le M} \int_0^M \phi_\pm(X,p+m,q)\, \tilde{f}_\pm(p+m,q)\, dq \ .$$

in $L_2(\Omega)$.

Finally, define linear operators

$$(6.41) \qquad \Phi_{\pm,p} : L_2(\Omega) \to \sum_{m\in Z} \oplus\, L_2(R_0)$$

by

$$(6.42) \qquad \Phi_{\pm,p}\, f = \{\tilde{f}_\pm(p+m,\cdot) \mid m \in Z\}$$

Then $\Phi_{+,p}$ and $\Phi_{-,p}$ are spectral mappings for A_p in the sense of

Theorem 6.7. For every bounded, Lebesgue-measurable function $\Psi(\lambda)$ defined on $p^2 \le \lambda < \infty$ one has

$$(\Phi_{\pm,p}\, \Psi(A_p) f)_m = \Psi(\omega^2(p+m,\cdot))(\Phi_{\pm,p} f)_m \ , \quad m \in Z \tag{6.43}$$

where $\Psi(A_p)$ is defined by the spectral theorem.

Finally, the orthogonality and completeness of the generalized eigenfunctions $\phi_\pm$ is expressed by

Theorem 6.8. The operators $\Phi_{+,p}$ and $\Phi_{-,p}$ are unitary.

It is clear from Parseval's relation (6.36) that $\Phi_{\pm,p}$ are isometries which proves the completeness relation

$$\Phi^*_{\pm,p}\, \Phi_{\pm,p} = 1 \ . \tag{6.44}$$

The surjectivity of $\Phi_{\pm,p}$ which is equivalent to the orthogonality relation

$$\Phi_{\pm,p}\, \Phi^*_{\pm,p} = 1 \tag{6.45}$$

is not a consequence of the spectral theorem. A proof of (6.45) by the method introduced in [30, p. 112ff] is given in §7 below.

Operators A_p that have Point Spectrum. It was shown in §4 that, in general, $\sigma_0(A_p)$ is discrete. Let $\mathcal{H}_0$ be the subspace of $L_2(\Omega)$ spanned by the eigenvectors of A_p and let $\dim \mathcal{H}_0 = N(p) - 1 \le \infty$. Let $\{\lambda_j(p) \mid 1 \le j < N(p)\}$ be the eigenvalues, repeated according to their multiplicity and enumerated so that $\lambda_j(p) \le \lambda_{j+1}(p)$. Let $\{\phi_j(X,p) \mid 1 \le j < N(p)\}$ be a corresponding orthonormal set of eigenfunctions.

Proceeding as before it is found that the diffracted plane waves $\phi_\pm(X,p+m,q)$ can be constructed and Theorem 5.1 holds with

$$E_{m,p} = \{q > 0 \mid \omega^2(p+m,q) \in T_p \cup \sigma_0(A_p)\} \tag{6.46}$$

which is still a countable set with no finite limit points. Similarly, the spectral family $\{\Pi_p(\mu)\}$ still satisfies (6.35) for $f \in L_2^{com}(\Omega)$ if $I \subset [p^2,\infty) - T_p - \sigma_0(A_p)$. It follows that $\Pi_p(\mu)$ differs from (6.39) only by the projection

$$\sum_{\lambda_j(p)\le\mu} \phi_j(X,p)\, \tilde{f}_j(p) \ , \quad \tilde{f}_j(p) = (\phi_j(\cdot,p),f)_{L_2(\Omega)} \tag{6.47}$$

and Parseval's relation and the eigenfunction expansion become

$$(6.48)\qquad \|f\|^2 = \sum_{j=1}^{N-1} |\tilde{f}_j(p)|^2 + \sum_{m\in Z} \|\tilde{f}_\pm(p+m,\cdot)\|^2 , \quad f \in L_2(\Omega) .$$

and

$$(6.49)\qquad f(X) = \sum_{j=1}^{N-1} \phi_j(X,p)\, \tilde{f}_j(p) + \sum_{m\in Z} \int_0^\infty \phi_\pm(X,p+m,q)\, \tilde{f}_\pm(p+m,q)\, dq ,$$

convergent in $L_2(\Omega)$. The form of the spectral family implies that A_p has no singular continuous spectrum: $L_2(\Omega) = \mathcal{H}_0 \oplus \mathcal{H}_{ac}$, where $\mathcal{H}_{ac}$ is the subspace of absolute continuity for A_p [13, Ch. X]. Finally, Theorem 6.8 must be modified to state that $\Phi_{+,p}$ and $\Phi_{-,p}$ are partial isometries with initial set $\mathcal{H}_{ac}$ and final set $\Sigma \oplus L_2(R_0)$:

$$(6.50)\qquad \Phi^*_{\pm,p}\, \Phi_{\pm,p} = P_{ac} , \quad \Phi_{\pm,p}\, \Phi^*_{\pm,p} = 1$$

where P_{ac} is the orthogonal projection of $L_2(\Omega)$ onto $\mathcal{H}_{ac}$.

§7. Proofs of the Results of §6

Theorem 6.1 is a direct consequence of Theorem 2.1 and the results of §4.

Proof of Lemma 6.2. The proof follows the plan of [30, Lemma 6.3]. Definitions (6.16) and (6.25) imply that if $f \in L_2^{com}(\Omega)$ then

$$(7.1)\qquad \begin{aligned} \tilde{f}(p+m,q,z) &= \int_{\mathrm{supp}\, f} \overline{\phi_0(X,p+m,q)}\, j(y)\, f(X)\, dX \\ &\quad + \int_{\mathrm{supp}\, f} \overline{\phi'(X,p+m,q,\overline{z})}\, f(X)\, dX \\ &= (J\, f)_0^\sim\, (p+m,q) + \int_{\mathrm{supp}\, f} \overline{R(A_p,\overline{z})\, M(\cdot,p+m,q)}\, f(X)\, dX \\ &= (J\, f)_0^\sim\, (p+m,q) + \int_{\Omega_{h,r}} \overline{M(X,p+m,q)}\, R(A_p,z)\, f(X)\, dX \\ &= (J\, f)_0^\sim\, (p+m,q) + \\ &\quad + \int_{\Omega_{h,r}} \overline{(\Delta+\omega^2(p+m,q))\{j(y)\phi_0(X,p+m,q)\}}\, R(A_p,z)\, f(X)\, dX \end{aligned}$$

since $\phi'(\cdot,p+m,q,z) = R(A_p,z)\, M(\cdot,p+m,q)$ by (6.13) and Theorem 4.8 and supp $M \subset \Omega_{h,r}$. The next-to-last equation follows from $R(A_p,z) = R(A_p,\overline{z})^*$. To derive (6.26) from (7.1) it is necessary to integrate by parts in the last integral. This cannot be done directly because $j(y)\ \phi_0(X,p+m,q) \notin L_2(\Omega)$. To complete the calculation introduce a function $\xi \in C^\infty(R)$ such that $\xi'(y) \le 0$, $\xi(y) = 1$ for $y \le 0$, $\xi(y) = 0$ for $y \ge 1$ and define

$$\xi_n(y) = \xi(y - n) = \begin{cases} 1 \ , \ y \le n \ , \\ 0 \ , \ y \ge n + 1 \ . \end{cases} \tag{7.2}$$

Then for $n \ge r$ one has $\xi_n(y) \equiv 1$ on $\Omega_{0,r}$ and hence

$$\begin{aligned} \tilde{f}(p+m,q,z) &= (J\ f)_0^{\sim}\ (p+m,q) \\ &+ \int_\Omega \overline{(\Delta+\omega^2(p+m,q))\{j(y)\phi_0(X,p+m,q)\}}\ \xi_n(y)\ R(A_p,z)\ f(X)\ dX \ . \end{aligned} \tag{7.3}$$

Now

$$j(y)\ \phi_0(X,p+m,q) \in D(A_p^{N,\ell oc}(\Omega))\ \ (\text{resp. } D(A_p^{D,\ell oc}(\Omega))\ . \tag{7.4}$$

This may be shown by interpreting $\exp\{-ipx\}\ \phi_0(X,p+m,q)$ as a function on the cylinder Ω^γ (see the proof of Lemma 4.1) and recalling that $j(y) = 0$ for $0 \le y \le (h+r)/2$. Moreover,

$$\xi_n\ R(A_p,z)f \in L_2^{1,p,com}(\Omega)\ \ (\text{resp. } L_2^{D,p,com}(\Omega)) \tag{7.5}$$

since $R(A_p,z)f \in D(A_p)$. Conditions (7.4), (7.5) and the integral identities of (3.19), (3.20) applied to $u = j\phi_0$ and $v = \xi_n\ R(A_p,z)f$ give

$$\begin{aligned} &\int_\Omega \Delta\ \overline{\{j(y)\phi_0(X,p+m,q)\}}\ \xi_n(y)\ R(A_p,z)\ f(X)\ dX \\ &= -\int_\Omega \nabla\ \overline{\{j(y)\phi_0(X,p+m,q)\}} \cdot \nabla\{\xi_n(y)\ R(A_p,z)\ f(X)\}\ dX \\ &= -\int_\Omega \nabla\ \overline{\{j(y)\xi_{n+1}(y)\phi_0(X,p+m,q)\}} \cdot \nabla\{\xi_n(y)\tilde{j}(y)R(A_p,z)f(X)\}\ dX \end{aligned} \tag{7.6}$$

where $\tilde{j} \in C_0^\infty(h,\infty)$ and $\tilde{j}(y) \equiv 1$ for $y \ge (h+r)/2$. Now

$$j(y)\ \xi_{n+1}(y)\ \phi_0(X,p+m,q) \in L_2^{1,p}(\Omega_{0,n+2}) \tag{7.7}$$

and

$$\xi_n(y)\ \tilde{j}(y)\ R(A_p,z)\ f(X) \in D(A_p^N(\Omega_{0,n+1})) \tag{7.8}$$

and a second application of the integral identity of (3.19), together with

(7.6), give

$$\text{(7.9)}\quad \begin{aligned} &\int_\Omega \Delta\,\overline{\{j(y)\phi_0(X,p+m,q)\}}\;\xi_n(y)\;R(A_p,z)\;f(X)\;dX \\ &= \int_\Omega \overline{j(y)\xi_{n+1}(y)\phi_0(X,p+m,q)}\;\Delta\{\xi_n(y)\tilde{j}(y)R(A_p,z)f(X)\}\;dX \\ &= \int_\Omega j(y)\;\overline{\phi_0(X,p+m,q)}\;\Delta\{\xi_n(y)R(A_p,z)f(X)\}\;dX \end{aligned}$$

because $\xi_{n+1}(y) \equiv 1$ on supp ξ_n and $\tilde{j}(y) \equiv 1$ on supp j. Also, Leibniz's rule for distribution derivatives implies

$$\text{(7.10)}\quad \Delta\{\xi_n R(A_p,z)f\} = \xi_n\;\Delta\;R(A_p,z)f + 2\xi_n'\;D_y R(A_p,z)f + \xi_n''\;R(A_p,z)f\;.$$

Combining this and the differential equation $\Delta R(A_p,z)f = -A_p R(A_p,z)f = -f - z\;R(A_p,z)f$ gives

$$\text{(7.11)}\quad \begin{aligned} (\Delta+\omega^2(p+m,q))\;\{\xi_n\;R(A_p,z)f\} &= -\xi_n f + (\omega^2-z)\xi_n\;R(A_p,z)f \\ &\quad + 2\xi_n'\;D_y R(A_p,z)f + \xi_n''\;R(A_p,z)f\;. \end{aligned}$$

Combining (7.3), (7.9) and (7.11) gives

$$\text{(7.12)}\quad \begin{aligned} \tilde{f}(p+m,q,z) &= (J\;f)_0^{\sim}\;(p+m,q) - \int_{\text{supp } f} \overline{\phi_0(X,p+m,q)}\;\xi_n(y)j(y)f(X)dX \\ &\quad + (\omega^2(p+m,q)-z)\int_\Omega \overline{\phi_0(X,p+m,q)}\;\xi_n(y)j(y)R(A_p,z)f(X)dX \\ &\quad + 2\int_\Omega \overline{\phi_0(X,p+m,q)}\;\xi_n'(y)j(y)D_y R(A_p,z)f(X)dX \\ &\quad + \int_\Omega \overline{\phi_0(X,p+m,q)}\;\xi_n''(y)\;j(y)\;R(A_p,z)f(X)dX\;. \end{aligned}$$

Now $\xi_n(y) \equiv 1$ on supp f and hence the first two terms of the right-hand side of (7.12) cancel for $n \geq n_0 = n_0(f)$. In view of the definition (3.28), (3.31) of the unitary spectral mapping $\Phi_{0,p}$ associated with $A_{0,p}$, equation (7.12) implies that for all $n \geq n_0$ one has

$$\text{(7.13)}\quad \begin{aligned} \tilde{f}(p+m,q,z) &= (\omega^2(p+m,q) - z)\;\{\Phi_{0,p}(\xi_n\;J\;R(A_p,z)f)\}_m(q) \\ &\quad + 2\;\{\Phi_{0,p}(\xi_n'\;J\;D_y R(A_p,z)f)\}_m(q) \\ &\quad + \{\Phi_{0,p}(\xi_n''\;J\;R(A_p,z)f)\}_m(q)\;. \end{aligned}$$

Now $J\ R(A_p,z)f \in L_2(B_0)$ and $J\ D_y\ R(A_p,z)f$ because $R(A_p,z)f \in D(A_p) \subset L_2^1(\Omega)$. Moreover, $0 \le \xi_n(y) \le 1$, $\xi_n(y) \to 1$ when $n \to \infty$ for all $y \ge 0$ and $\text{supp}\ \xi_n' \cup \text{supp}\ \xi_n'' \subset \{y \mid n \le y \le n+1\}$. It follows by Lebesgue's dominated convergence theorem that $\xi_n\ J\ R(A_p,z)f \to J\ R(A_p,z)f$, $\xi_n'\ J\ D_y\ R(A_p,z)f \to 0$ and $\xi_n''\ J\ R(A_p,z)f \to 0$ in $L_2(B_0)$ when $n \to \infty$. Hence passage to the limit $n \to \infty$ in (7.13) gives

$$\tilde{f}(p+m,q,z) = (\omega^2(p+m,q) - z)\ \{\Phi_{0,p}(J\ R(A_p,z)f)\}_m(q) \tag{7.14}$$

which is equivalent to (6.26).

Proof of Lemma 6.3. This result follows from the continuity of $\tilde{f}(p+m,q,\lambda\pm i\sigma)$ for $q > 0$, $\lambda \in [p^2,\infty) - T_p$ and $\sigma \ge 0$. The details of the proof are precisely the same as in [30, Lemma 6.6] and are therefore not repeated here.

Proof of Lemma 6.4. The starting point for the proof of (6.34) is equation (7.1) with $z = \lambda + i\sigma$, $\lambda \in I \subset [p^2,\infty) - T_p$ and $0 < \sigma \le \sigma_0$. (7.1) can be written

$$\tilde{f}(p+m,q,z) = (J\ f)_0^{\sim}\ (p+m,q) + g(p+m,q,z) \tag{7.15}$$

where

$$g(p+m,q,z) = \int_{\Omega_{h,r}} \overline{M(X,p+m,q)}\ R(A_p,z)\ f(X)\ dX\ . \tag{7.16}$$

Note that (see (6.9))

$$M(X,p+m,q) = 2\ D_y\ \{j'(y)\ \phi_0(X,p+m,q)\} - j''(y)\ \phi_0(X,p+m,q) \tag{7.17}$$

and hence

$$g(p+m,q,z) = g_1(p+m,q,z) + g_2(p+m,q,z) \tag{7.18}$$

where

$$g_1(p+m,q,z) = -\int_{\Omega_{h,r}} \overline{\phi_0(X,p+m,q)}\ j''(y)\ R(A_p,z)\ f(X)\ dX \tag{7.19}$$

and

$$(7.20)\quad g_2(p+m,q,z) = 2\int_{\Omega_{h,r}} D_y \{j'(y)\ \overline{\phi_0(X,p+m,q)}\}\ R(A_p,z)\ f(X)\ dX\ .$$

In the last integral note that $R(A_p,z)f$ is in $L_2^{2,\ell oc}([h,\infty),L_2(-\pi,\pi))$ (cf. Lemma 4.1) while $j'(y)\ \phi_0(X,p+m,q) \in C_0^\infty([h,\infty),L_2(-\pi,\pi))$ and $j(r) = 0$. It follows that

$$(7.21)\quad g_2(p+m,q,z) = -2\int_{\Omega_{h,r}} \overline{\phi_0(X,p+m,q)}\ j'(y)\ D_y\ R(A_p,z)\ f(X)\ dX\ .$$

Note that (7.19) and (7.21) extend by continuity to $z = \lambda \pm i\sigma$, with $\lambda \in I$ and $0 \le \sigma \le \sigma_0$, by Theorem 4.15.

Equations (7.15) and (7.18) imply that

$$(7.22)\quad \begin{aligned} |\tilde{f}(p+m,q,z)|^2 \le 4(|(J\ f)_0^\sim\ (p+m,q)|^2 &+ |g_1(p+m,q,z)|^2 \\ &+ |g_2(p+m,q,z)|^2)\ . \end{aligned}$$

Moreover, Parseval's relation (3.29) for $A_{0,p}$ implies

$$(7.23)\quad \sum_{m\in Z}\int_0^\infty |(J\ f)_0^\sim\ (p+m,q)|^2\ dq = \|J\ f\|^2_{L_2(B_0)} \le \|f\|^2_{L_2(\Omega_{0,k})}$$

where supp $f \subset \Omega_{0,k}$. Hence to prove Lemma 6.4 it will suffice to prove (6.34) with $\tilde{f}$ replaced by g_1 and g_2. For g_1, equation (7.19), Parseval's relation (3.29) and Corollary 4.17 imply

$$(7.24)\quad \begin{aligned} \sum_{m\in Z}\int_0^\infty |g_1(p+m,q,z)|^2\ dq &= \|j''\ R(A_p,z)f\|^2_{L_2(B_0)} \\ &\le (\text{Max}\ |j''(y)|)^2\ \|R(A_p,z)f\|^2_{L_2(\Omega_{0,r})} \\ &\le (\text{Max}\ |j''(y)|)^2\ C^2\ \|f\|^2_{L_2(\Omega_{0,k})} \end{aligned}$$

for all $z = \lambda \pm i\sigma$ with $\lambda \in I$ and $\sigma \in [0,\sigma_0]$ where $C = C(I,p,\sigma_0,k,r)$ $= C(I,p,\sigma_0,f)$ is the constant of Corollary 4.17. The proof of Lemma 6.4 may be completed by noting that the integral (7.21) for g_2 has the same form as (7.19) but with $j''\ R(A_p,z)f$ replaced by $2\ j'\ D_y\ R(A_p,z)f$. An estimate for g_2 of the same form as (7.24) follows because the $L_2(\Omega_{0,r})$ norm of $D_y\ R(A_p,z)f$ is majorized by the $L_2^1(\Delta,\Omega_{0,r})$ norm of $R(A_p,z)f$.

<u>Proofs of Theorems 6.5, 6.6 and 6.7</u>. These results all follow from (6.35) by the spectral theorem and standard Hilbert space methods and

therefore will not be given here. A detailed development of these arguments in the case of exterior domains may be found in [30, pp. 109ff].

Proof of Theorem 6.8. Only the orthogonality relation (6.45) need be proved. The proof presented here is based on a method introduced in [30] for the case of exterior domains. The proof for the case of grating domains differs in some important technical details from that of [30] and is therefore presented in full here.

The isometry $\Phi_{\pm,p}$ is known to satisfy (6.45) if and only if [30, p. 116]

$$N(\Phi^*_{\pm,p}) = \{0\} \ ; \tag{7.25}$$

i.e., the null space of $\Phi^*_{\pm,p}$ contains only the zero vector. Equation (6.45) will be proved by verifying (7.25). The following two lemmas are needed.

Lemma 7.1. For all $h = \{h_m(q)\} \in \Sigma \oplus L_2(R_0)$ one has

$$\Phi^*_{\pm,p}\, h(X) = \underset{M\to\infty}{\ell.\text{i.m.}} \sum_{|m|\le m} \int_0^M \phi_\pm(X,p+m,q)\, h_m(q)\, dq \tag{7.26}$$

where the convergence is in $L_2(\Omega)$.

Lemma 7.2. Let $h \in N(\Phi^*_{\pm,p})$ and let $\Psi(\lambda)$ be a bounded Lebesgue measurable function on $\lambda \ge p^2$. Then

$$h' = \{\Psi(\omega^2(p+m,q))h_m(q)\} \in N(\Phi^*_{\pm,p}) \ . \tag{7.27}$$

Proofs of Lemmas 7.1 and 7.2. Lemma 7.1 is a direct consequence of (6.38) and (6.42); see [30, Lemma 6.17]. To prove Lemma 7.2 let $f \in L_2(\Omega)$ and note that the definitions of $\Phi_{\pm,p}$ and $\Phi^*_{\pm,p}$ and Theorem 6.7 imply

$$\begin{aligned}(f,\Phi^*_{\pm,p}\, h') &= (\Phi_{\pm,p}\, f,h') \\ &= \sum_{m\in Z} \int_0^\infty \overline{\tilde f_\pm(p+m,q)\ \Psi(\omega^2(p+m,q))}\ h_m(q)\ dq \\ &= \sum_{m\in Z} \int_0^\infty \overline{\overline{\Psi}(\omega^2(p+m,q))\ \tilde f_\pm(p+m,q)}\ h_m(q)\ dq \\ &= \sum_{m\in Z} \int_0^\infty \overline{(\Phi_{\pm,p}\ \overline{\Psi}(A_p)f)_m\ (q)}\ h_m(q)\ dq \\ &= (\Phi_{\pm,p}\ \overline{\Psi}(A_p)f,h) = (\overline{\Psi}(A_p)f,\Phi^*_{\pm,p}h) = 0 \ .\end{aligned} \tag{7.28}$$

This proves (7.27) since $f \in L_2(\Omega)$ is arbitrary.

Choice of $\Psi(\lambda)$. Let

$$(7.29) \qquad I = [a,b] \subset [p^2,\infty) - T_p$$

and define

$$(7.30) \qquad \Psi(\lambda) = \exp\{-it\,\lambda^{1/2}\}\;\chi_I(\lambda)\;,\quad \lambda \geq p^2\;,$$

where $t \in R$ and $\chi_I(\lambda)$ is the characteristic function of I. It will be shown that Lemma 7.2 with this class of functions $\Psi(\lambda)$ implies (7.25). The following notation will be used.

$$(7.31) \qquad N = \{m : \omega^2(p+m,q) \in I \text{ for some } q > 0\}\;.$$

Note that N is a finite set. Moreover, $q \to \omega^2(p+m,q)$ is monotone for $q \in R_0$ and hence for each $m \in N$

$$(7.32) \qquad \lambda = \omega^2(p+m,q) \in I \Leftrightarrow q = \sqrt{\lambda - (p+m)^2} \in I_{m,p} \subset R_0 - E_{m,p}$$

where $I_{m,p}$ is a compact interval and $E_{m,p}$ is defined by (6.15). With this choice of Ψ, Lemmas 7.1 and 7.2 imply that if $h \in N(\Phi^*_{\pm,p})$ then

$$(7.33) \qquad \begin{aligned} \Phi^*_{\pm,p}\, h'(X) &= \sum_{m\in Z} \int_0^\infty \phi_\pm(X,p+m,q)\;\Psi(\omega^2(p+m,q))\; h_m(q)\; dq \\ &= \sum_{m\in Z} \int_{I_{m,p}} \phi_\pm(X,p+m,q)\; e^{-it\omega(p+m,q)}\; h_m(q)\; dq = 0 \end{aligned}$$

in $L_2(\Omega)$. The left hand side of (7.33) defines a solution of the d'Alembert equation in Ω. Its behavior for $t \to \mp\infty$ will be determined and shown to imply (7.25). For this purpose one needs the

Far-Field Form of $\phi_\pm(X,p+m,q)$. This phrase means the form of $\phi_\pm(x,y,p+m,q)$ for large y; i.e., far from the grating. To derive it note that (6.14) and Lemma 4.1 imply that

$$(7.34) \qquad \phi'_\pm(X,p+m,q) = \sum_{\ell\in Z} \phi'_{\pm\ell}(y,p+m,q)\;\exp\{i(p+\ell)x\}$$

in $L_2^{2,\ell oc}(\Omega_h)$. Moreover, for $y \geq r$

$$(7.35) \qquad \phi'_{\pm\ell}(y,p+m,q) = a^\pm_\ell(p+m,q)\;\exp\{i\,y\,w_{p+\ell}(\omega^2(p+m,q) \pm i0)\}\;.$$

It follows that for $q \in I_{m,p}$ and $X \in \Omega_r$

(7.36)
$$\phi_{\pm}(X,p+m,q) = \phi_0(X,p+m,q) + \sum_{\ell \in L} a_\ell^{\pm}(p+m,q) \exp\{i\,x\,p_\ell \pm i\,y\,q_\ell\} + \rho_{\pm}(X,p+m,q)$$

where

(7.37)
$$L = L(p,I) = \{\ell : |p+\ell| < \omega(p+m,q)\} ,$$

and

(7.38)
$$(p_\ell, q_\ell) = (p+\ell, (\omega^2(p+m,q) - (p+\ell)^2)^{1/2})$$

while

(7.39)
$$\rho_{\pm}(X,p+m,q) = \sum_{\ell \in L'} \phi'_{\pm\ell}(y,p+m,q) \exp\{i(p+\ell)x\}$$

where

(7.40)
$$L' = L'(p,I) = \{\ell : |p+\ell| > \omega(p+m,q)\} .$$

It is important to note that for $q \in I_{m,p}$ the sets L and L' are independent of q and depend on p and I only. An estimate for the term $\rho_{\pm}$ in (7.36) is given by

<u>Lemma 7.3</u>. There exists a constant $\mu = \mu(p,I) > 0$ and for each $r' > r$ a constant $C = C(I,p,m,r,r')$ such that

(7.41)
$$|\rho_{\pm}(X,p+m,q)| \le C\, e^{-\mu y} \text{ for } X \in \Omega_{r'} , \quad q \in I_{m,p} .$$

<u>Proof of Lemma 7.3</u>. For brevity write $u(X) = \phi'_{\pm}(X,p+m,q)$ and note that $u \in F_{p,\zeta,r}$ with $\zeta = \omega^2(p+m,q) \pm i0 \in \overline{M_p^{\mp}} - \Sigma_p$. In particular by Lemma 4.1

(7.42)
$$u(X) = \sum_{\ell \in Z} u_\ell(y) \exp\{i(p+\ell)x\} \text{ in } L_2^{2,\ell oc}(\Omega_h)$$

and

(7.43)
$$u_\ell(y) = u_\ell(y') \exp\{-(y-y')((p+\ell)^2 - \omega^2(p+m,q))^{1/2}\}$$

for all $y,y' \geq r$ and all $\ell \in L'$. Now by a Sobolev inequality [1, p. 32] there exists a $C_0 = C_0(h,r)$ such that

$$|u_\ell(r)|^2 \leq C_0^2 \left(\int_h^r |u'(y)|^2 \, dy + \int_h^r |u_\ell(y)|^2 \, dy\right) . \tag{7.44}$$

Moreover,

$$\|u(\cdot,y)\|^2_{L_2(-\pi,\pi)} = 2\pi \sum_{\ell\in Z} |u_\ell(y)|^2 , \tag{7.45}$$

$$\|D_y \, u(\cdot,y)\|^2_{L_2(-\pi,\pi)} = 2\pi \sum_{\ell\in Z} |u'(y)|^2 , \tag{7.46}$$

which, with (7.44) imply

$$\begin{aligned} |u_\ell(r)|^2 &\leq C_0^2(2\pi)^{-1}(\|D_y \, u\|^2_{h,r} + \|u\|^2_{h,r}) \\ &\leq C_0^2(2\pi)^{-1} \, \|u\|^2_{1;h,r} ; \end{aligned} \tag{7.47}$$

i.e.,

$$|\phi'_{\pm\ell}(r,p+m,q)|^2 \leq C_0^2(2\pi)^{-1} \, \|\phi'_\pm(\cdot,p+m,q)\|^2_{1;h,r} . \tag{7.48}$$

Now the right hand side of (7.48) is a continuous function of $q \in R_0 - E_{m,p}$ by Theorem 6.1. Thus there exists a constant $C_1 = C_1(I,p,m,r)$ such that

$$|\phi'_{\pm\ell}(r,p+m,q)| \leq C_1 \text{ for all } q \in I_{m,p} . \tag{7.49}$$

Next, recalling (7.29), define

$$\mu = \mu(p,I) = \operatorname*{Min}_{\ell\in L'} \{(p+\ell)^2 - b^2\}^{1/2} \tag{7.50}$$

so that for all $q \in I_{m,p}$ and $\ell \in L'(p,I)$ one has

$$\{(p+\ell)^2 - \omega^2(p+m,q)\}^{1/2} \geq \{(p+\ell)^2 - b^2\}^{1/2} \geq \mu > 0 . \tag{7.51}$$

Then for $r' > r$ and $X \in \Omega_{r'}$, $q \in I_{m,p}$ one has the estimates

$$\begin{aligned} |\rho_\pm(X,p+m,q)| &\leq \sum_{\ell\in L'} |\phi'_{\pm\ell}(y,p+m,q)| \\ &\leq \sum_{\ell\in L'} |\phi'_{\pm\ell}(r,p+m,q)| \exp\{-(y-r)((p+\ell)^2-(\omega^2(p+m,q))^{1/2}\} \end{aligned} \tag{7.52}$$

$$\leq C_1 \sum_{\ell \in L'} \exp\{-(y-r)\{(p+\ell)^2 - b^2\}^{1/2}\}$$

(7.52 cont.)

$$\leq \left(C_1 \sum_{\ell \in L'} \exp\{-(r'-r)\{(p+\ell)^2 - b^2\}^{1/2}\}\right) \exp\{-\mu(y-r')\}$$

which implies (7.41).

Proof of Theorem 6.8 (continued). Substitution of the far-field form (7.36) for $\phi_\pm$ in the identity (7.33) gives the identity

$$u_0(t,X) + u_1(t,X) + u_2(t,X) = 0 \text{ in } L_2(\Omega) \tag{7.53}$$

for all $t \in R$ where

$$u_0(t,X) = \sum_{m\in N} \int_{I_{m,p}} \phi_0(X,p+m,q)\, e^{-it\omega(p+m,q)}\, h_m(q)\, dq \tag{7.54}$$

$$u_1(t,X) = \sum_{m\in N} \int_{I_{m,p}} \left(\sum_{\ell\in L} a_\ell^\pm(p+m,q) e^{i(xp_\ell \pm yq_\ell)} \right) e^{-it\omega(p+m,q)}\, h_m(q)\, dq \tag{7.55}$$

$$u_2(t,X) = \sum_{m\in N} \int_{I_{m,p}} \rho_\pm(X,p+m,q)\, e^{-it\omega(p+m,q)}\, h_m(q)\, dq\ . \tag{7.56}$$

Note that $u_0(t,X)$ has an extension to $X \in B_0$ such that (see (3.32), (3.33))

$$u_0(t,\cdot) = \exp\{-it\, A_{0,p}^{1/2}\}\, h_I^\circ \tag{7.57}$$

where

$$h_I^\circ = \Phi_{0,p}^* \{\chi_{m,p}\, h_m : m \in Z\} \in L_2(B_0) \tag{7.58}$$

and $\chi_{m,p}$ is the characteristic function of $I_{m,p}$. In particular, one has

$$\|u_0(t,\cdot)\|^2_{L_2(B_0)} = \|h_I^\circ\|^2_{L_2(B_0)} = \sum_{m\in N} \int_{I_{m,p}} |h_m(q)|^2\, dq\ . \tag{7.59}$$

The proof of Theorem 6.8 will be completed by showing that

$$\lim_{t\to\mp\infty} \|u_0(t,\cdot)\|_{L_2(B_0)} = 0\ . \tag{7.60}$$

It follows from (7.59), (7.60) that $h_m(q) = 0$ for almost all $q \in I_{m,p}$. But $\lambda = \omega^2(p+m,q)$ maps $R_0 - E_{m,p}$ bijectively onto $[p^2,\infty) - T_p$ (see (6.15)).

Thus given any $m \in Z$ and any interval $I_{m,p} \subset R_0 - E_{m,p}$ there is an interval $I \subset [p^2,\infty) - T_p$ such that the above relations hold. Thus $h_m(q) \equiv 0$ in $R_0 - E_{m,p}$ for every $m \in Z$, whence $h = 0$ in $\Sigma \oplus L_2(R_0)$ which prove (7.25).

Proof of (7.60). Consider first the function $u_1(t,X)$ defined by (7.55). It can be written

$$u_1(t,X) = \sum_{m\in N} u_{1,m}(t,X) \tag{7.61}$$

where

$$u_{1,m}(t,X) = \sum_{\ell\in L} u_{1,m,\ell}(t,y) \exp\{i(p+\ell)x\} \tag{7.62}$$

and

$$u_{1,m,\ell}(t,y) = \int_{I_{m,p}} a_\ell^{\pm}(p+m,q)\, e^{\pm iyq_\ell - it\omega(p+m,q)}\, h_m(q)\, dq \, . \tag{7.63}$$

In the last integral

$$\begin{aligned} q_\ell &= \{\omega^2(p+m,q) - (p+\ell)^2\}^{1/2} \\ &= \{q^2 + (p+m)^2 - (p+\ell)^2\}^{1/2} \equiv Q(q,p+m,p+\ell) \, . \end{aligned} \tag{7.64}$$

Make the change of variable

$$q' = q_\ell = Q(q,p+m,p+\ell) \tag{7.65}$$

in (7.63). Since

$$\omega^2(p+m,q) = \omega^2(p+\ell,q') \tag{7.66}$$

one has

$$q = Q(q',p+\ell,p+m) \tag{7.67}$$

and

$$u_{1,m,\ell}(t,y) = \int_{I'_{m,\ell,p}} a_\ell^{\pm}(p+m,q)\, e^{\pm iyq' - it\omega(p+\ell,q')}\, h_m(q)\, \frac{\partial q}{\partial q'}\, dq' \, . \tag{7.68}$$

Now each of these integrals has the form of a modal wave in a simple waveguide [31, §5]. Moreover, it was shown in [31] that

$$(7.69)\qquad \lim_{t\to\mp\infty} \|u_{1,m,\ell}(t,\cdot)\|_{L_2(R_0)} = 0 .$$

Thus it follows from (7.69),

$$(7.70)\qquad \|u_{1,m}(t,\cdot,y)\|^2_{L_2(-\pi,\pi)} = 2\pi \sum_{\ell\in L} |u_{1,m,\ell}(t,y)|^2 ,$$

$$(7.71)\qquad \|u_{1,m}(t,\cdot)\|^2_{L_2(B_0)} = 2\pi \sum_{\ell\in L} \|u_{1,m,\ell}(t,\cdot)\|^2_{L_2(R_0)} ,$$

and

$$(7.72)\qquad \|u_1(t,\cdot)\|_{L_2(B_0)} \le \sum_{m\in N} \|u_{1,m}(t,\cdot)\|_{L_2(B_0)}$$

that

$$(7.73)\qquad \lim_{t\to\mp\infty} \|u_1(t,\cdot)\|_{L_2(B_0)} = 0 .$$

It will be shown next that the function $u_2(t,X)$ defined by (7.56) satisfies

$$(7.74)\qquad \lim_{t\to\mp\infty} \|u_2(t,\cdot)\|_{L_2(\Omega)} = 0 .$$

This is a consequence of the following two lemmas.

Lemma 7.4. The function $u(t,X) = u_2(t,X)$ defined by (7.56) has the properties

$$(7.75)\qquad u(t,\cdot) \in L_2(\Omega) \text{ for all } t \in R ,$$

$$(7.76)\qquad \lim_{t\to\pm\infty} \|u(t,\cdot)\|_{0,k} = 0 \text{ for all } k > r ,$$

and there exists a $\mu > 0$ and for each $r' > h$ a constant $C = C(r')$ such that

$$(7.77)\qquad |u(t,X)| \le C\, e^{-\mu y} \text{ for all } X \in \Omega_{r'} \text{ and } t \in R .$$

Lemma 7.5. If $u(t,X)$ is any function having properties (7.75), (7.76), (7.77) then

$$(7.78) \qquad \lim_{t\to\pm\infty} \|u(t,\cdot)\|_{L_2(\Omega)} = 0 .$$

Proof of Lemma 7.4. To verify (7.75) note that by (7.53), $u(t,\cdot) = u_2(t,\cdot) = -u_0(t,\cdot) - u_1(t,\cdot)$ in $L_2(\Omega)$. But $u_0(t,\cdot) \in L_2(B_0)$ by the spectral theory of $A_{0,p}$ (§3 above) and $u_1(t,\cdot) \in L_2([-\pi,\pi] \times R)$ by the theory of waveguides as developed in [31]. Thus the restrictions of these functions to Ω are in $L_2(\Omega)$.

The decomposition $u = u_2 = -u_0 - u_1$ also implies (7.76) because u_0 and u_1 both represent waves in simple waveguides which have this local decay property; see [31].

Property (7.77) is a consequence of the definition of u_2, equation (7.56), and Lemma 7.3. Indeed, combining (7.41) and (7.46) gives (7.77) with $\mu = \mu(p,I)$ defined by (7.50) and

$$(7.79) \qquad C = C(I,p,m,r,r') \sum_{M\in N} \int_{I_{m,p}} |h_m(q)| \, dq .$$

Proof of Lemma 7.5. Conditions (7.75) and (7.77) imply that one has for each $r' > h$ and $k > r'$,

$$(7.80) \qquad \begin{aligned} \|u(t,\cdot)\|^2_{L_2(\Omega)} &= \|u(t,\cdot)\|^2_{0,k} + \|u(t,\cdot)\|^2_{k,\infty} \\ &= \|u(t,\cdot)\|^2_{0,k} + \int_k^\infty \int_{-\pi}^{\pi} |u(t,x,y)|^2 \, dxdy \\ &\le \|u(t,\cdot)\|^2_{0,k} + C^2 \int_k^\infty \int_{-\pi}^{\pi} e^{-2\mu y} \, dxdy \\ &= \|u(t,\cdot)\|^2_{0,k} + (\pi C^2/\mu) \, e^{-2\mu k} \end{aligned}$$

where $C = C(r')$ is independent of k. Making $t \to \pm\infty$ in (7.80) with k fixed gives, by (7.76),

$$(7.81) \qquad \limsup_{t\to\pm\infty} \|u(t,\cdot)\|^2_{L_2(\Omega)} \le (\pi C^2/\mu) \, e^{-2\mu k}$$

for all $k > r'$. This implies (7.78) since the left hand side of (7.81) is independent of k.

Proof of Theorem 6.8 (concluded). The proof may be concluded by verifying (7.60). Now the identity (7.53) implies

$$(7.82)\qquad \begin{aligned}\|u_0(t,\cdot)\|_{L_2(B_0)} &\le \|u_0(t,\cdot)\|_{L_2(B_0-\Omega)} + \|u_0(t,\cdot)\|_{L_2(\Omega)}\\ &\le \|u_0(t,\cdot)\|_{L_2(B_0-\Omega)} + \|u_1(t,\cdot)\|_{L_2(B_0)} + \|u_2(t,\cdot)\|_{L_2(\Omega)} .\end{aligned}$$

Moreover, $B_0 - \Omega$ is bounded and hence $u_0(t,\cdot) \to 0$ in $L_2(B_0 - \Omega)$ by the local decay property for $A_{0,p}$. The remaining terms on the right hand side of (7.82) tend to zero when $t \to \mp \infty$ by (7.73) and (7.74).

§8. The Rayleigh-Bloch Wave Expansions for A.

This section presents a construction, based on the results of §6, of the R-B diffracted plane wave eigenfunctions $\psi_\pm(X,p,q)$ and a derivation of the corresponding R-B wave expansions for A. For brevity the derivation is restricted to the cases for which A has no surface waves; that is, $\sigma_0(A_p) = \phi$ for all p. The modifications that are needed when there are surface waves are indicated at the end of the section.

In this section $\psi_{0\pm}(X,p,q)$ denotes the R-B wave eigenfunction for A_0. The defining properties of $\psi_\pm(X,p,q)$ can then be written

$$(8.1)\qquad \psi_\pm(\cdot,p,q) \in D(A^{\ell oc}) , \quad (p,q) \in R_0^2 ,$$

$$(8.2)\qquad (\Delta + \omega^2(p,q))\, \psi_\pm(X,p,q) = 0 \text{ in } G ,$$

$$(8.3)\qquad \psi_\pm(X,p,q) = \psi_{0\pm}(X,p,q) + \psi'_\pm(X,p,q) \text{ in } R_h^2 ,$$

where ψ'_+ (resp., ψ'_-) is an outgoing (resp., incoming) R-B wave for G.

The construction of $\psi_\pm$ will be based on the discussion at the end of §3. Thus if $(p,q) \in R_0^2$ and $p = p_0 + m$ where $p_0 \in (-1/2,1/2]$ and $m \in Z$ then the functions $\psi_\pm(X,p,q)$ are defined by

$$(8.4)\qquad \psi_\pm(X,p,q) = \mathcal{O}^{p_0} \phi_\pm(X,p_0+m,q) ,$$

or, more explicitly,

$$(8.5)\qquad \psi_\pm(x,y,p,q) = \exp\{2\pi i \ell p_0\}\, \phi_\pm(x-2\pi\ell,y,p_0+m,q) , \quad (x,y) \in \Omega^{(\ell)} .$$

Theorem 6.1 then implies

<u>Theorem 8.1</u>. Let G be a grating domain of the class defined in §1 and let $A = A(G)$ have no surface waves. Then there exist unique R-B diffracted

plane waves $\psi_\pm(X,p,q)$ for each $(p,q) \in R_0^2 - E$, where E is the exceptional set (2.30). Moreover, the mapping $(p,q) \to \psi_\pm(\cdot,p,q) \in L_2^{1,\ell oc}(\Delta,G)$ is continuous for $(p,q) \in R_0^2 - E$.

The principal step in the proof of Theorem 8.1 is to show that $\psi_\pm$, defined piece-wise by (8.5), satisfies (8.1). This may be done by a simple distribution-theoretic calculation based on the p-periodic boundary condition for $\phi_\pm$. Details are given in §9 below. The uniqueness statement follows from Theorem 2.1 since $\sigma_0(A_p) = \phi$ for all p is assumed.

The R-B wave expansions for A will now be derived from the eigenfunction expansions for A_p of §6. The first step is to establish Parseval's relation for A. The special case of functions $f \in L_2^{com}(G)$ is treated first.

Theorem 8.2. For all $f \in L_2^{com}(G)$ define

$$(8.6) \qquad \hat{f}_\pm(p,q) = \int_G \overline{\psi_\pm(X,p,q)}\, f(X)\, dX\,, \quad (p,q) \in R_0^2 - E\,.$$

Then

$$(8.7) \qquad \hat{f}_\pm \in C(R_0^2 - E) \cap L_2(R_0^2)\,, \text{ and}$$

$$(8.8) \qquad \|f\|_{L_2(G)} = \|\hat{f}_\pm\|_{L_2(R_0^2)}\,.$$

Proof. The finiteness of $\hat{f}_\pm(p,q)$ for $(p,q) \in R_0^2 - E$ and the property $\hat{f}_\pm \in C(R_0^2 - E)$ follow from the last statement of Theorem 8.1. To establish the rest of the theorem note the following identity for functions $f \in L^{com}(G)$ and points $(p,q) \in R_0^2 - E$.

$$(8.9) \qquad \begin{aligned} \hat{f}_\pm(p,q) &= \int_G \overline{\psi_\pm(X,p,q)}\, f(X)\, dX = \sum_{\ell\in Z} \int_{\Omega(\ell)} \overline{\psi_\pm(X,p,q)}\, f(X)\, dX \\ &= \sum_{\ell\in Z} \int_\Omega \overline{\psi_\pm(x+2\pi\ell,y,p,q)}\, f(x+2\pi\ell,y)\, dxdy \\ &= \sum_{\ell\in Z} \int_\Omega \overline{\phi_\pm(x,y,p,q)\, e^{-2\pi i\ell p}}\, f(x+2\pi\ell,y)\, dxdy \\ &= \int_\Omega \overline{\phi_\pm(x,y,p,q)} \left\{ \sum_{\ell\in Z} e^{-2\pi i\ell p}\, f(x+2\pi\ell,y) \right\} dxdy \\ &= \int_\Omega \overline{\phi_\pm(x,y,p,q)}\, F(x,y,p)\, dxdy \end{aligned}$$

where

$$(8.10) \qquad F(x,y,p) = \sum_{\ell\in Z} e^{-2\pi i\ell p} f(x+2\pi\ell,y) , \quad (x,y) \in \Omega .$$

Notice that all the sums in (8.9) are finite when $f \in L_2^{com}(G)$. Moreover, (8.10) is a Fourier series in p with a fixed finite number of non-zero terms for all $(x,y) \in \Omega$.

Equation (8.9) establishes a relation between the eigenfunction expansions for A and A_p. Thus replacing p in (8.9) by p + m with $p \in (-1/2,1/2]$ and $m \in Z$ one has

$$(8.11) \qquad \hat{f}_\pm(p+m,q) = \tilde{F}_\pm(p+m,q,p)$$

in the notation of §6. In particular, (8.11) and Parseval's relation for A_p, applied to $F(\cdot,p)$, give

$$(8.12) \qquad \int_\Omega |F(X,p)|^2 \, dX = \sum_{m\in Z} \int_0^\infty |\hat{f}_\pm(p+m,q)|^2 \, dq .$$

Noting the continuity of $p \to F(\cdot,p) \in L_2(\Omega)$ and integrating (8.12) over $p \in (-1/2,1/2]$ gives

$$(8.13) \qquad \begin{aligned} \int_{-1/2}^{1/2} \int_\Omega |F(X,p)|^2 \, dX\, dp &= \sum_{m\in Z} \int_{-1/2}^{1/2} \int_0^\infty |\hat{f}_\pm(p+m,q)|^2 \, dq dp \\ &= \int_{R_0^2} |\hat{f}_\pm(p,q)|^2 \, dp dq = \|\hat{f}_\pm\|^2_{L_2(R_0^2)} . \end{aligned}$$

In particular, $\hat{f}_\pm \in L_2(R_0^2)$ which completes the proof of (8.7). To verify (8.8) note that Parseval's formula for Fourier series implies that

$$(8.14) \qquad \int_{-1/2}^{1/2} |F(X,p)|^2 \, dp = \sum_{\ell\in Z} |f(x+2\pi\ell,y)|^2 , \quad X \in \Omega ,$$

where the sum has a fixed finite number of terms for all $X \in \Omega$. Integrating (8.14) over $X \in \Omega$ and applying Fubini's theorem gives

$$(8.15) \qquad \begin{aligned} \int_{-1/2}^{1/2} \int_\Omega |F(X,p)|^2 \, dX\, dp &= \sum_{\ell\in Z} \int_\Omega |f(x+2\pi\ell,y)|^2 \, dX \\ &= \sum_{\ell\in Z} \int_{\Omega(\ell)} |f(X)|^2 \, dX = \int_G |f(X)|^2 \, dX = \|f\|^2_{L_2(G)} . \end{aligned}$$

Combining (8.13) and (8.15) gives (8.8).

The extension of Parseval's relation to all $f \in L_2(G)$ follows from Theorem 8.2 by a standard technique using the denseness of $L_2^{com}(G)$ in $L_2(G)$. Thus, writing

$$G_M = G \cap \{X \mid x^2 + y^2 < M^2\} \,, \tag{8.16}$$

one has

Corollary 8.3. The limits

$$\hat{f}_\pm(p,q) = \underset{M\to\infty}{\ell.\text{i.m.}} \int_{G_M} \overline{\psi_\pm(X,p,q)}\, f(X)\, dX \tag{8.17}$$

exist in $L_2(R_0^2)$ and Parseval's relation (8.8) holds for all $f \in L_2(G)$.

A representation of the spectral family $\{\Pi(\mu) \mid \mu \geq 0\}$ of the grating propagator A will now be derived from Corollary 8.3. The key fact is described by

Theorem 8.4. The resolvent $R(A,z) = (A - z)^{-1}$ of the grating propagator A satisfies the relation

$$\|R(A,z)f\|^2_{L_2(G)} = \int_{R_0^2} \frac{|\hat{f}_\pm(p,q)|^2}{|\omega^2(p,q)-z|^2}\, dpdq \tag{8.18}$$

for all $f \in L_2(G)$ and all $z \in C - [0,\infty)$.

To prove Theorem 8.4 it is enough to verify (8.18) for all $f \in L_2^{com}(G)$. The idea for doing this is to define

$$u(X) = R(A,z)\, f(X) \tag{8.19}$$

and to apply Parseval's relation to $v_M = \phi_M u$ where $\phi_M \in C_0^2(R^2)$. For a suitable choice of ϕ_M one has

$$v_M = R(A,z)(f + g_M) \tag{8.20}$$

where

$$g_M = -2\nabla u \cdot \nabla\phi_M - u\, \Delta\phi_M \tag{8.21}$$

and

(8.22) $$\hat{v}_{M\pm}(p,q) = (\hat{f}_\pm(p,q) + \hat{g}_{M\pm}(p,q))/\omega^2(p,q) - z$$

whence

(8.23) $$\|\phi_M\, R(A,z)f\| = \|(\hat{f}_\pm + \hat{g}_{M\pm})/(\omega^2 - z)\| .$$

Passage to the limit $M \to \infty$ then gives (8.18). For the case of the Dirichlet boundary condition one may take $\phi_M(X) = \psi(|X| - M)$ where $\psi \in C^\infty(R)$ satisfies $\psi(\tau) \equiv 1$ for $\tau \leq 0$ and $\psi(\tau) \equiv 0$ for $\tau \geq 1$. For the case of the Neumann boundary condition ϕ_M must be chosen more carefully, using the condition $G \in S$, to ensure that v_M satisfy the boundary condition. The details of the construction are given in §9.

The R-B wave expansions for A follow easily from Corollary 8.3 and Theorem 8.4. They are formulated as

Theorem 8.5. For all $f \in L_2(G)$ the spectral family $\{\Pi(\mu) \mid \mu \geq 0\}$ of A satisfies

(8.24) $$\Pi(\mu)\ f(X) = \int_{D_\mu} \psi_\pm(X,p,q)\ \hat{f}_\pm(p,q)\ dpdq$$

where

(8.25) $$D_\mu = R_0^2 \cap \{(p,q) \mid p^2 + q^2 \leq \mu\} .$$

In particular, every $f \in L_2(G)$ has the R-B wave expansion

(8.26) $$f(X) = \underset{M\to\infty}{\ell.i.m.} \int_{D_M} \psi_\pm(X,p,q)\ \hat{f}_\pm(p,q)\ dpdq .$$

The relation (8.24) is a direct consequence of the relation

(8.27) $$\|\Pi(I)f\|^2_{L_2(G)} = \int_{R_0^2} \chi_I(\omega^2(p,q))\ |\hat{f}_\pm(p,q)|^2\ dpdq$$

where I is a subinterval of $[0,\infty)$ with characteristic function χ_I. (8.27) follows easily from (8.18) and Stone's formula. Note that (8.27) implies the absolute continuity of the grating propagators.

To formulate the orthogonality and completeness relations for the R-B wave expansions define linear operators

(8.28) $$\Phi_\pm : L_2(G) \to L_2(R_0^2)$$

by

$$\Phi_{\pm} f = \hat{f}_{\pm} . \tag{8.29}$$

Then Φ_+ and Φ_- are spectral mappings for A in the sense of

Theorem 8.6. For every bounded, Lebesgue-measurable function $\Psi(\lambda)$ defined on $0 \leq \lambda < \infty$

$$\Phi_{\pm} \Psi(A) = \Psi(\omega^2(\cdot)) \Phi_{\pm} \tag{8.30}$$

where $\Psi(A)$ is defined by the spectral theorem.

Moreover, one has

Theorem 8.7. The R-B wave expansions are orthogonal and complete in the sense that Φ_+ and Φ_- are unitary operators:

$$\Phi_{\pm}^* \Phi_{\pm} = 1 \text{ and } \Phi_{\pm} \Phi_{\pm}^* = 1 . \tag{8.31}$$

Relations (8.30) and the completeness relation $\Phi_{\pm}^* \Phi_{\pm} = 1$ follow easily from the spectral theorem. The orthogonality relation $\Phi_{\pm} \Phi_{\pm}^* = 1$ can be deduced from the corresponding property of $\Phi_{\pm p}$, Theorem 6.8. Indeed, it is sufficient to prove that

$$(\Phi_{\pm} \Phi_{\pm}^* f - f, f) = 0 \tag{8.32}$$

for all f in a dense subset of $L_2(R_0^2)$. This may be verified by direct calculation using $f \in C_0^\infty(R_0^2 - E)$ and the orthogonality relation for $\Phi_{\pm p}$. The details are given in §9.

Operators A that Admit R-B Surface Waves. It was shown in §2 that for each $p \in (-1/2, 1/2]$ A may have R-B surface waves $\psi_j(X,p)$ and eigenvalues $\lambda_j(p)$ with x-momentum p. The functions $\phi_j(X,p) = \psi_j(X,p)|_{\Omega}$ are precisely the eigenfunctions of A_p. The principal difficulty in constructing an eigenfunction expansion for A in this case is in constructing families of R-B surface waves $\psi_j(X,p)$ and eigenvalues $\lambda_j(p)$ whose dependence on p is sufficiently regular. The "axiom of choice" definition (independent choice for each p) is inadequate to give even measurability in p. This was pointed out in the author's paper on the analogous, but simpler, case of Bloch waves in crystals [32].

If ∂G is a union of smooth curves (class C^3) then the Green's functions (4.44), (4.45) can be used to construct an integral equation for the eigenfunctions $\phi_j(X,p)$. In this case the method of [32] can be used to construct "almost holomorphic" families $\{\phi_j(X,p)\}$.

In the general case there is a one-to-one correspondence between eigenfunctions $\phi_j(X,p)$ of A_p and eigenfunctions $\theta_j(X,p)$ of $A_{p,\zeta,r}$ with eigenvalues $\pi_p(\zeta) \in [p^2,\infty)$ given by $\theta_j(\cdot,p) = P_{p,\zeta,r}\,\phi_j(\cdot,p)$. The eigenvalues of $A_{p,\zeta,r}$ are isolated, with finite multiplicity, and may be studied by the methods of analytic perturbation theory (Kato [13, Ch. 7]). These problems will not be pursued here.

If a sufficiently regular family of R-B surface waves for A has been constructed the eigenfunction expansions for A may be derived by the method introduced above. Thus, defining $\psi_j(X,p) \equiv 0$ when $j \geq N(p)$, equation (8.12) must be replaced by

$$(8.33)\qquad \int_\Omega |F(X,p)|^2\, dX = \sum_{j=1}^{\infty} |\hat{f}_j(p)|^2 + \sum_{m\in Z} \int_0^\infty |\hat{f}_\pm(p+m,q)|^2\, dq$$

where

$$(8.34)\qquad \hat{f}_j(p) = \int_G \overline{\psi_j(X,p)}\, f(X)\, dX\ .$$

Integration over $p \in (-1/2,1/2]$ gives the Parseval relation

$$(8.35)\qquad \|f\|^2_{L_2(G)} = \sum_{j=1}^{\infty} \|\hat{f}_j\|^2_{L_2(-1/2,1/2)} + \|\hat{f}_\pm\|^2_{L_2(R_0^2)}\ .$$

The corresponding representation of the spectral family is

$$(8.36)\qquad \Pi(\mu)\, f(X) = \int_{-1/2}^{1/2} \sum_{\lambda_j(p)\leq\mu} \psi_j(X,p)\, \hat{f}_j(p)\, dp + \int_{D_\mu} \psi_\pm(X,p,q)\, \hat{f}_\pm(p,q)\, dpdq\ .$$

§9. Proofs of the Results of §8.

Proof of Theorem 8.1. It will be shown that if $\phi_\pm(X,p+m,q)$ are the generalized eigenfunctions for A_p whose existence is guaranteed by Theorem 6.1 then the functions $\psi_\pm(X,p,q)$ defined by (8.5) have properties (8.1), (8.2), (8.3). This will prove the existence statement of Theorem 8.1. Note that $q \in E_{m,p} \Leftrightarrow (p,q) \in E$ (see (2.30) and (6.15)). Hence the construction (8.5) is valid for $(p,q) \in R_0^2 - E$.

The sets $D(A^{\ell oc})$ are characterized in the cases of the Neumann and Dirichlet boundary conditions by (see (1.26), (1.28))

$$(9.1)\quad D(A^{N,\ell oc}(G)) = L_2^{1,\ell oc}(\Delta,G) \cap \{u : (1.14) \text{ holds for } v \in L_2^{1,com}(G)\} ,$$

$$(9.2)\qquad D(A^{D,\ell oc}(G)) = L_2^{1,\ell oc}(\Delta,G) \cap L_2^{D,\ell oc}(G) .$$

As a first step it will be verified that (8.5) defines a function $\psi_\pm(\cdot,p,q) \in L_2^{1,\ell oc}(\Delta,G)$ for each $(p,q) \in R_0^2 - E$. It is clear that $\psi_\pm(\cdot,p,q) \in L_2^{\ell oc}(G) \subset \mathcal{D}'(G)$ for $(p,q) \in R_0^2 - E$ because $\phi_\pm(\cdot,p_0+m,q) \in L_2^{\ell oc}(\Omega)$ for $p_0 \in (-1/2,1/2]$, $m \in Z$ and $q \in R_0 - E_{m,p}$. It remains to show that $\nabla\psi_\pm(\cdot,p,q)$ and $\Delta\psi_\pm(\cdot,p,q)$, as elements of $\mathcal{D}'(G)$, are also in $L_2^{\ell oc}(G)$. Now by definition $\phi_\pm(\cdot,p+m,q) \in L_2^{1,\ell oc}(\Delta,\Omega)$ and hence (8.5) implies

$$(9.3)\qquad \psi_\pm(\cdot,p,q) \in L_2^{1,\ell oc}\left(\Delta, \bigcup_{\ell\in Z} \Omega^{(\ell)}\right) .$$

Hence, it is only necessary to verify that $\psi_\pm(\cdot,p,q)$, $\nabla\psi_\pm$ and $\Delta\psi_\pm$ are locally square integrable near the lines $\{(2\ell+1)\pi \times \gamma : \ell \in Z\}$ (see (3.4)). Moreover, $\psi_\pm((2\ell+1)\pi \pm 0,y,p,q)$ and $D_1\psi_\pm((2\ell+1)\pi \pm 0,y,p,q)$ exist in $L_2^{\ell oc}(\gamma)$ (see the discussion preceding (3.7)) and the p-periodic boundary condition for $\phi_\pm$ and (8.5) imply

$$(9.4)\qquad \begin{cases} \psi_\pm((2\ell+1)\pi+0,\cdot,p,q) = \psi_\pm((2\ell+1)\pi-0,\cdot,p,q) \\ D_1\psi_\pm((2\ell+1)\pi+0,\cdot,p,q) = D_1\psi_\pm((2\ell+1)\pi-0,\cdot,p,q) . \end{cases}$$

The proof that $\psi_\pm(\cdot,p,q) \in L_2^{1,\ell oc}(\Delta,G)$ will be completed by proving

<u>Lemma 9.1</u>. The distribution derivatives $D_j\psi_\pm(\cdot,p,q)$ are given by

$$(9.5)\quad D_j\psi_\pm(x,y,p,q) = \exp\{2\pi i\ell p_0\}\, D_j\phi_\pm(x-2\pi\ell,y,p_0+m,q) , \quad (x,y) \in \Omega^{(\ell)} ,$$

for $j = 1,2$. Moreover, $\psi_\pm(\cdot,p,q)$ satisfies (8.2) as a distribution on G.

<u>Proof of Lemma 9.1</u>. (9.5) will be proved for $j = 1$. Thus it will be shown that for all $\theta \in C_0^\infty(G)$ one has

$$(9.6)\qquad \int_G \psi_\pm\, D_1\theta\, dX = -\int_G D_1\psi_\pm\, \theta\, dX$$

where $D_1\psi_\pm \in L_2^{\ell oc}(G)$ is defined by (9.5). This will be verified for functions θ with $\text{supp}\, \theta \subset \Omega^{(0)} \cup \Omega^{(1)} \cup (\pi \times \gamma)$. In this case (9.6) is a consequence of (9.4) and the equations

$$(9.7)\qquad \int_{\Omega^{(0)}} \psi_\pm\, D_1\theta\, dX = -\int_{\Omega^{(0)}} D_1\psi_\pm\, \theta\, dX + \int_\gamma \psi_\pm(\pi-0,y,p,q)\, dy$$

$$(9.8)\qquad \int_{\Omega^{(1)}} \psi_\pm\, D_1\theta\, dX = -\int_{\Omega^{(1)}} D_1\psi_\pm\, \theta\, dX - \int_\gamma \psi_\pm(\pi+0,y,p,q)\, dy\ .$$

Equation (9.7) may be verified by calculating

$$(9.9)\qquad \int_{\Omega^{(0)}} \psi_\pm\, D_1(\phi_\delta\theta)\, dX\ ,$$

where $\phi_\delta(x) = \phi((x-\pi)/\delta)$, $\phi_\delta(x) \equiv 1$ for $x \le \pi - \delta$, $\phi_\delta(x) \equiv 0$ for $x \ge \pi$ and $0 \le \phi_\delta(x) \le 1$, and then making $\delta \to 0$. The technique is explained in [31, p. 57ff]. The case of a general $\theta \in C_0^\infty(G)$ may be proved in the same way. The proof of (9.5) for $j = 2$ is similar. Moreover, an analogous calculation, based on (9.4), gives

$$(9.10)\quad \Delta\psi_\pm(x,y,p,q) = \exp\{2\pi i\ell p_0\}\, \Delta\phi_\pm(x-2\pi\ell,y,p_0+m,q)\ ,\quad (x,y) \in \Omega^{(\ell)}\ ,$$

and it follows from (6.2) that $\psi_\pm$ satisfies (8.2).

<u>Proof of Theorem 8.1 (continued)</u>. To complete the proof that $\psi_\pm(\cdot,p,q) \in D(A^{\ell oc})$ in the Neumann case, condition (1.14) must be proved for $v \in L^{1,com}(G)$. Now for such a v one has, by Lemma 9.1,

$$(9.11)\qquad \begin{aligned} \int_G \Delta\psi_\pm\, \overline{v}\, dX &= \sum_{\ell\in Z} \int_{\Omega^{(\ell)}} \Delta\psi_\pm\, \overline{v}\, dX \\ &= \sum_{\ell\in Z} \int_\Omega \Delta\psi_\pm(x+2\pi\ell,y,p,q)\, \overline{v}\,(x+2\pi\ell,y)\, dX \\ &= \sum_{\ell\in Z} \int_\Omega \Delta\phi_\pm(x,y,p,q)\, e^{2\pi i\ell p}\, \overline{v}\,(x+2\pi\ell,y)\, dX \\ &= \int_\Omega \Delta\phi_\pm\, \overline{u}\, dX \end{aligned}$$

where

$$(9.12)\qquad u(x,y) = \sum_{\ell\in Z} e^{-2\pi i\ell p}\, v(x+2\pi\ell,y) \in L_2^{1,p,com}(\Omega)\ .$$

Note that the sums in (9.11), (9.12) are finite because $v \in L_2^{1,com}(G)$. A similar calculation gives

$$(9.13)\qquad \int_G \nabla\psi_\pm \cdot \nabla\overline{v}\, dX = \int_\Omega \Delta\phi_\pm \cdot \overline{\nabla u}\, dX$$

and adding (9.11), (9.13) gives

$$(9.14)\qquad \int_G \{\Delta\psi_\pm\, \overline{v} + \nabla\psi_\pm \cdot \nabla\overline{v}\}\, dX = \int_\Omega \{\Delta\phi_\pm\, \overline{u} + \nabla\phi_\pm \cdot \overline{\nabla u}\}\, dX = 0$$

because $\phi_\pm \in D(A^{N,\ell oc}(\Omega))$ and $u \in L_2^{1,p,com}(\Omega)$ (see (3.19)).

To complete the proof that $\psi_\pm(\cdot,p,q) \in D(A^{\ell oc})$ in the Dirichlet case, it must be shown that $\psi_\pm(\cdot,p,q) \in L_2^{D,\ell oc}(G)$ = Closure of $C_0^\infty(G)$ in $L_2^{1,\ell oc}(G)$. This follows immediately from (8.5) because $\psi_\pm(\cdot,p,q)$ is p-periodic and $\phi_\pm(\cdot,p+m,q) \in L_2^{D,p,\ell oc}(\Omega)$ = Closure of $C_p^\infty(\Omega)$ in $L_2^{1,\ell oc}(\Omega)$. To see this note that on any set $K \cap G$ where K is compact in R^2 the functions $\theta \in C_p^\infty(G)$ coincide with functions $\theta' = \phi\theta$ where $\phi \in C_0^\infty(R^2)$ and $\phi(X) \equiv 1$ on K.

It has been shown that $\psi_\pm(\cdot,p,q)$, defined for all $(p,q) \in R_0^2 - E$ by (8.4), satisfies (8.1) and (8.2). Condition (8.3) is also immediate because ψ_0 and ϕ_0 satisfy $\psi_0(X,p,q) = O^{p_0}\, \phi_0(X,p_0+m,q)$ (see (3.27)) and hence

$$(9.15)\qquad \psi_\pm'(X,p,q) = O^{p_0}\, \phi_\pm'(X,p_0+m,q)\ .$$

It follows that ψ_+' (resp., ψ_-') is an outgoing (resp., incoming) R-B wave for G.

The uniqueness of $\psi_\pm(\cdot,p,q)$ was proved in §8. To complete the proof of Theorem 8.1 the continuity of $(p,q) \to \psi_\pm(\cdot,p,q) \in L_2^{1,\ell oc}(\Delta,G)$ for $(p,q) \in R_0^2 - E$ must be shown. Note that since $\psi_\pm$ satisfies (8.2) it will be enough to prove the continuity of the mapping $(p,q) \to \psi_\pm(\cdot,p,q) \in L_2^{1,\ell oc}(G)$. Thus it must be shown that for each compact $K \subset R^2$ and each $(p_0,q_0) \in R_0^2 - E$ one has

$$(9.16)\qquad \begin{cases} \|\psi_\pm(\cdot,p,q) - \psi_\pm(\cdot,p_0,q_0)\|_{L_2(K\cap G)} \to 0 \\ \|\nabla\psi_\pm(\cdot,p,q) - \nabla\psi_\pm(\cdot,p_0,q_0)\|_{L_2(K\cap G)} \to 0 \end{cases}$$

when $(p,q) \to (p_0,q_0)$. For the functions $\psi_0(\cdot,p,q)$ the continuity conditions (9.16) follow from (1.33), (1.34) by direct calculation. For $\psi_\pm'(\cdot,p,q)$ they follow from (8.5) and the continuity of $(p,q) \to \phi_\pm'(\cdot,p,q) \in L_2^{1,\ell oc}(\Omega)$: i.e.,

$$(9.17)\qquad \begin{cases} \|\phi_\pm'(\cdot,p,q) - \phi_\pm'(\cdot,p_0,q_0)\|_{L_2(K\cap\Omega)} \to 0 \\ \|\nabla\phi_\pm'(\cdot,p,q) - \nabla\phi_\pm'(\cdot,p_0,q_0)\|_{L_2(K\cap\Omega)} \to 0 \end{cases}$$

when $(p,q) \to (p_0,q_0)$. (9.17) is a consequence of Theorem 4.15 and the definitions (6.13) and (6.14). (9.17) and (8.5) imply (9.16) because $K \cap G$ is contained in a finite union of the sets $K \cap \Omega^{(\ell)}$.

Proof of Theorem 8.2. This was given in §8.

Proof of Corollary 8.3. As remarked in §8, these results follow from Theorem 8.2 and the fact that $L_2^{com}(G)$ is dense in $L_2(G)$. The details may be found in [30, p. 109] where the corresponding results are proved for exterior domains.

Proof of Theorem 8.4. The proof outlined in §8 will be completed here. The two boundary conditions will be discussed separately.

The Dirichlet Case. Proceeding as in §8, let $f \in L_2^{com}(G)$ and define

$$u = R(A^D(G),z)f \tag{9.18}$$

and

$$v_M(X) = \phi_M(X)\ u(X)\ , \quad x \in G\ , \tag{9.19}$$

where $\phi_M(X) = \psi(|X| - M) \in C_0^2(R^2)$ satisfies $\phi_M(X) \equiv 1$ on G_M and supp $\phi_M \subset G_{M+1}$. Then it is easy to verify that $v_M \in D(A^D(G))$ and

$$\begin{aligned}(A^D(G)-z)\ v_M(X) &= -(\Delta+z)\ \phi_M(X)\ u(X) \\ &= \phi_M(X)\ f(X) - 2\nabla u \cdot \nabla\phi_M - u\ \Delta\phi_M \\ &= f(X) + g_M(X) \text{ for } M \geq M_0(f)\ ,\end{aligned} \tag{9.20}$$

where g_M is defined by (8.21), because $\phi_M(X) \equiv 1$ on supp f for $M \geq M_0(f)$.

Equation (9.20) implies (8.20). To verify (8.22) note that by (8.2) one has

$$\begin{aligned}\hat{v}_{M\pm}(p,q) &= \int_G \overline{\psi_\pm(X,p,q)}\ v_M(X)\ dX \\ &= -\omega^{-2}(p,q) \int_G \Delta\ \overline{\psi_\pm(X,p,q)}\ v_M(X)\ dX\ .\end{aligned} \tag{9.21}$$

Now $\psi_\pm(\cdot,p,q) \in L_2^{1,\ell oc}(\Delta,G)$ and hence

$$(9.22) \qquad \int_G \{(\Delta\overline{\psi}_\pm)\, v_M + \nabla\overline{\psi}_\pm \cdot \nabla v_M\}\, dX = 0$$

because $v_M \in L_2^D(G)$ and supp v_M is compact. Indeed, $v_M = \lim \phi_n$ in $L_2^1(G)$ where $\phi_n \in C_0^\infty(G)$ and (9.22) holds with v_M replaced by ϕ_n by the distribution definitions of $\Delta\psi_\pm$ and $\nabla\psi_\pm$. Similarly, one has

$$(9.23) \qquad \int_G \{\overline{\psi}_\pm(\Delta\, v_M) + \nabla\overline{\psi}_\pm \cdot \nabla v_M\}\, dX = 0$$

because $v_M \in D(A^D(G))$, supp v_M is compact and $\psi_\pm \in L_2^{D,\ell oc}(G)$. Combining (9.21), (9.22) and (9.23) gives

$$(9.24) \qquad \hat{v}_{M\pm}(p,q) = -\omega^{-2}(p,q) \int_G \overline{\psi_\pm(X,p,q)}\, \Delta v_M(X)\, dX .$$

Finally, combining (9.20) and (9.24) gives

$$(9.25) \qquad \begin{aligned} \hat{v}_{M_\pm}(p,q) &= -\omega^{-2}(p,q)(\psi_\pm(\cdot,p,q),\Delta v_M) \\ &= \omega^{-2}(p,q)(\psi_\pm(\cdot,p,q), f+g_M+zv_M) \\ &= \omega^{-2}(p,q)(\hat{f}_\pm(p,q) + \hat{g}_{M\pm}(p,q) + z\hat{v}_{M\pm}(p,q)) . \end{aligned}$$

Solving this equation for $\hat{v}_{M_\pm}$ gives (8.22) and hence (8.23).

To find the limiting form of (8.23) for $M \to \infty$ note that $|\phi_M(X)| \le 1$ and $\phi_M(X) \to 1$ for all $X \in G$ when $M \to \infty$. Moreover,

$$(9.26) \qquad \lim_{M\to\infty} g_M = 0 \text{ in } L_2(G)$$

because in the definition (8.21) ∇u and u are in $L_2(G)$, $|\nabla\phi_M(I)|$ and $\Delta\phi_M(X)$ are bounded uniformly for all M and supp $g_M \subset G_{M+1} - G_M$. Hence passage to the limit $M \to \infty$ in (8.23) gives (8.18) for $f \in L_2^{com}(G)$. The general case follows by a density argument.

<u>The Neumann Case</u>. The method presented above can be used. However, the definition of the multiplier ϕ_M must be modified to ensure that $v_M \in D(A^N(G))$. If $\phi_M \in C_0^2(\overline{G})$ then it is easy to show that $v_M \in L_2^1(\Delta,G)$. The hypothesis $G \in S$ of §1 will be used to construct a function $\phi_M \in C_0^2(\overline{G})$ such that $v_M = \phi_M u$ also satisfies the Neumann boundary condition. The construction is similar to the one used above to prove Theorem 4.6 in the Neumann case.

To construct ϕ_M let $\sigma(x,y)$, $\tau(x,y)$ be the tangent-normal coordinates defined in the neighborhoods $\mathcal{O} + (2\pi m,0)$ of the points $((2m-1)\pi,y_0)$ as in §5 following (5.177). Define ξ_2 by (5.178) as before and let $\eta_1,\eta_3 \in C^2(R)$ satisfy $0 \le \eta_j(\alpha) \le 1$ and

$$(9.27) \qquad \eta_j(\alpha) = \begin{cases} 1 \text{ for } \alpha \le -\delta_j \\ 0 \text{ for } \alpha \ge \delta_j \end{cases}$$

where $\delta_j > 0$. Define

$$(9.28) \qquad \phi_M^1(x,y) = \eta_1(\sigma)\ \xi_2(\tau) + \eta_3(x - (2M+1)\pi)[1 - \xi_2(\tau)]$$

for all $(x,y) \in G \cap \{(x,y) : x \ge 0\}$. Note that if $0 < \delta < \pi$ then for δ_1, δ_2, δ_3 small enough one has

$$(9.29) \qquad \phi_M^1(x,y) = \begin{cases} 1 \text{ for } x \le (2M+1)\pi - \delta\ , \\ 0 \text{ for } x \ge (2M+1)\pi + \delta\ . \end{cases}$$

Extend ϕ_M^1 to the rest of G by

$$(9.30) \qquad \phi_M^1(x,y) = 1 - \phi_M^1(-x,y) \text{ for } (x,y) \in G \cap \{(x,y) : x \le 0\}\ .$$

Finally let $\phi_M^2(y) \in C^2(R)$ satisfy $0 \le \phi_M^2(y) \le 1$, $\phi_M^2(y) \equiv 1$ for $y \le M$, $\phi_M^2(y) \equiv 0$ for $y \ge M + 1$ and define

$$(9.31) \qquad \phi_M(x,y) = \phi_M^1(x,y)\ \phi_M^2(y)\ .$$

Then ϕ_M has the desired properties. It is clear that $\phi_M \in C_0^2(\overline{G})$ and

$$(9.32) \quad \operatorname{supp} \phi_M \subset \{(x,y) : -(2M+1)\pi - \delta \le x \le (2M+1)\pi + \delta, 0 \le y \le M+1\}\ .$$

Moreover, in the strip $|x - (2M+1)\pi| \le \delta$, $0 \le y \le h$, one has $\xi_2(\tau) \equiv 1$ and hence $\phi_M(x,y) = \eta_1(\sigma(x,y))$. Similarly, in $|x + (2M+1)\pi| \le \delta$, $0 \le y \le h$ one has $\phi_M(x,y) = 1 - \eta_1(\sigma(x,y))$. This property implies that $v_M = \phi_M u$ satisfies the Neumann boundary condition on Γ; see (5.184). The remainder of the proof of Theorem 8.4 is the same as in the Dirichlet case.

<u>Proof of Theorem 8.5</u>. It was remarked in §8 that (8.24) and (8.26) are direct consequences of (8.27) (see [30, p. 110]). Relation (8.27) will be

derived from Theorem 8.4 and Stone's formula. The latter states that if $I = [a,b] \subset R$ then for all $f \in L_2(G)$ one has

$$\frac{1}{2}(f,[\Pi(b)+\Pi(b-)-\Pi(a)-\Pi(a-)]f) = \lim_{\sigma\to 0+} \frac{\sigma}{\pi}\int_I \|R(A,\lambda+i\sigma)f\|^2\, d\lambda \, . \tag{9.33}$$

Now Theorem 8.4 and Fubini's theorem imply that

$$\begin{aligned} \frac{\sigma}{\pi}\int_I \|R(A,\lambda+i\sigma)f\|^2\, d\lambda &= \frac{\sigma}{\pi}\int_I\int_{R_0^2} \frac{|\hat{f}_\pm(p,q)|^2}{|\omega^2(p,q)-\lambda-i\sigma|^2}\, dpdqd\lambda \\ &= \int_{R_0^2}\left(\frac{\sigma}{\pi}\int_I \frac{d\lambda}{(\lambda-\omega^2(p,q))^2+\sigma^2}\right) |\hat{f}_\pm(p,q)|^2\, dpdq \, . \end{aligned} \tag{9.34}$$

Moreover, if

$$K(\sigma,p,q) = \frac{\sigma}{\pi}\int_I \frac{d\lambda}{(\lambda-\omega^2(p,q))^2+\sigma^2} \tag{9.35}$$

then $0 \le K(\sigma,p,q) \le 1$ for all $(p,q) \in R_0^2$ and $\sigma > 0$ and $\lim K(\sigma,p,q) = \chi_I(\omega^2(p,q))$ for $\sigma \to 0$; [34, p. 98]. Hence (9.33) and (9.34) imply

$$\frac{1}{2}(f,[\Pi(b)+\Pi(b-)-\Pi(a)-\Pi(a-)]f) = \int_{R_0^2} \chi_I(\omega^2(p,q))\ |\hat{f}_\pm(p,q)|^2\, dpdq \tag{9.36}$$

by Lebesgue's dominated convergence theorem. On making $a \to b$ in (9.36) and using the relation $\Pi((b-)-) = \Pi(b-)$ one finds that $\Pi(b) = \Pi(b-)$ for all $b \in R$. Then putting $\Pi(b-) = \Pi(b)$, $\Pi(a-) = \Pi(a)$ in (9.36) gives (8.27).

Proof of Theorem 8.6. This result can be proved by the method used for the case of exterior domains in [30, p. 113]. The multiplier ϕ_m of [30, p. 114] may be replaced by the function ϕ_M used to prove Theorem 8.4. The remaining details are the same as in [34] and will not be repeated here.

Proof of Theorem 8.7. It will suffice to prove the relation (8.32), or equivalently

$$\|\Phi_\pm^* f\| = \|f\| \, , \tag{9.37}$$

for all $f \in C_0^\infty(R_0^2 - E)$.

As a first step, note that for all $f(p,q) \in L_2(R_0^2)$ one has

$$(\Phi_\pm^* f)(X) = \underset{M\to\infty}{\ell.i.m.} \int_{D_M} \psi_\pm(X,p,q)\, f(p,q)\, dpdq \, . \tag{9.38}$$

The simple proof is the same as for the case of exterior domains [30, p. 117]. If $f \in C_0^\infty(R_0^2 - E)$ then (9.38) can be written

$$(9.39)\qquad \begin{aligned}(\Phi^* f)(X) &= \int_{R_0^2} \psi_\pm(X,p,q)\, f(p,q)\, dpdq \\ &= \sum_{m\in Z} \int_{B_0} \psi_\pm(X,p+m,q)\, f(p+m,q)\, dpdq\end{aligned}$$

and only a finite number of terms in the sum are non-zero. In particular, the definition (8.5) of $\psi_\pm$ implies that

$$(9.40)\qquad (\Phi_\pm^* f)(X) = \sum_{m\in Z} \int_{B_0} e^{2\pi i\ell p}\, \phi_\pm(x-2\pi\ell,y,p+m,q)\, f(p+m)\, dpdq$$

for $X \in \Omega^{(\ell)}$.

Next note that

$$(9.41)\qquad \begin{aligned}\|\Phi^* f\|^2 &= \int_G |\Phi_\pm^* f(X)|^2\, dX = \sum_{\ell\in Z} \int_{\Omega^{(\ell)}} |\Phi_\pm^*\, f(X)|^2\, dX \\ &= \sum_{\ell\in Z} \int_\Omega |\Phi_\pm^* f(x+2\pi\ell,y)|^2\, dxdy\ .\end{aligned}$$

Now (9.40) implies that for $(x,y) \in \Omega$

$$(9.42)\qquad \begin{aligned}(\Phi_\pm^* f)(x+2\pi\ell,y) &= \sum_{m\in Z} \int_{-1/2}^{1/2} \left[\int_0^\infty e^{2\pi i\ell p}\, \phi_\pm(x,y,p+m,q)\, f(p+m,q)dq\right] dp \\ &= \int_{-1/2}^{1/2} e^{2\pi i\ell p}\, (\Phi_{\pm,p}^*\, \{f(p+\cdot,\cdot)\})(X)\, dp\end{aligned}$$

by Lemma 7.1. The interchange of summation and integration is elementary because the sum is finite for $f \in C_0^\infty(R_0^2 - E)$. Equation (9.42) states that the left hand side of the equation, as a function of $\ell \in Z$, is the set of Fourier coefficients of the function of p defined by $\Phi_{\pm,p}^*\ \{f(p+\cdot,\cdot)\}$. Thus Parseval's relation for Fourier series implies

$$(9.43)\qquad \sum_{\ell\in Z} |\Phi_\pm^* f(x+2\pi\ell,y)|^2 = \int_{-1/2}^{1/2} |(\Phi_{\pm,p}^*\, \{f(p+\cdot,\cdot)\})(X)|^2\, dp\ .$$

Integrating (9.43) over $X \in \Omega$ and using (9.41) gives, again by Fubini's theorem,

$$\|\Phi_{\pm}^* f\|^2 = \int_{-1/2}^{1/2} \int_{\Omega} |(\Phi_{\pm,p}^* \{f(p+\cdot,\cdot)\})(X)|^2 \, dX \, dp$$

(9.44)

$$= \int_{-1/2}^{1/2} \|\Phi_{\pm,p}^* \{f(p+\cdot,\cdot)\}\|^2 \, dp \, .$$

Now the orthogonality property for $\Phi_{\pm,p}^*$, Theorem 6.8, implies that

$$\|\Phi_{\pm,p}^* \{f(p+\cdot,\cdot)\}\|^2 = \|\{f(p+\cdot,\cdot)\}\|^2$$

(9.45)

$$= \sum_{m \in Z} \int_0^\infty |f(p+m,q)|^2 \, dq \, .$$

Combining (9.44) and (9.45) gives

(9.46) $$\|\Phi_{\pm}^* f\|^2 = \sum_{m \in Z} \|f\|^2_{L_2(B_0+(m,o))} = \|f\|^2_{L_2(R_0^2)}$$

which is equivalent to (9.37).

§10. The Initial-Boundary Value Problem for the Scattered Fields.

The goal of the remaining sections of this monograph is to analyze the structure of the transient acoustic and electromagnetic fields in grating domains G that are generated by sources which are localized in space and time. If the sources act during an interval $T \leq t \leq 0$ then the wave field may be described by a real-valued scalar function $u(t,X)$ that is a solution of the initial-boundary value problem

(10.1) $$D_t^2 u - \Delta u = 0 \text{ for all } t > 0 \text{ and } X \in G \, ,$$

(10.2) $$D_\nu u = 0 \text{ (resp., } u = 0) \text{ for all } t \geq 0 \text{ and } X \in \partial G \, ,$$

(10.3) $$u(0,X) = f(X) \text{ and } D_t u(0,X) = g(X) \text{ for all } X \in G \, .$$

The functions f, g which characterize the sources of the field will be assumed to have compact supports in G.

The initial-boundary value problem in its classical formulation (10.1) -(10.3) will have a solution only if ∂G and f and g are sufficiently smooth. However, for arbitrary domains G, if the initial state has finite energy; i.e.,

$$\int_G \{|\nabla f(X)|^2 + |g(X)|^2\}\, dX < \infty , \tag{10.4}$$

then the problem (10.1)-(10.3) has a unique generalized solution with finite energy (= solution wFE). This result was proved in [28]. In this section the results of §8 are used to derive an R-B wave representation of the solution wFE.

Consider the grating propagator of §1:

$$A = A^N(G) \text{ (resp., } A^D(G)) . \tag{10.5}$$

For arbitrary domains G, A is a selfadjoint realization in $\mathcal{H} = L_2(G)$ of the operator $-\Delta$ with the Neumann (resp., Dirichlet) boundary condition. Moreover, $A \geq 0$ and Kato's theory of sesquilinear forms [13] implies that $D(A^{1/2}) = L_2^1(G)$ (resp., $L_2^D(G)$). It follows that if

$$f \in L_2^1(G) \text{ (resp., } L_2^D(G)) \text{ and } g \in L_2(G) \tag{10.6}$$

then (10.4) holds and

$$u(t,\cdot) = (\cos t\, A^{1/2})f + (A^{-1/2} \sin t\, A^{1/2})g \tag{10.7}$$

is the unique solution wFE of (10.1)-(10.3) [30]. In particular,

$$u \in C^1(R, L_2(G)) \cap C(R, D(A^{1/2})) \tag{10.8}$$

and the initial conditions hold in $L_2(G)$. The boundary conditions are incorporated in the definition of D(A) and of solution wFE. The d'Alembert equation (10.1) holds in a suitable weak form [30]. The scattered fields studied below are the solutions wFE defined by (10.6), (10.7).

It was shown in [30, Ch. 3] that solutions wFE in arbitrary domains have a representation

$$u(t,X) = \operatorname{Re}\{v(t,X)\} , \quad v(t,\cdot) = e^{-itA^{1/2}} h \tag{10.9}$$

provided that f and g satisfy (10.6) and $g \in D(A^{-1/2})$. The complex-valued function $h \in D(A^{1/2})$ is related to the initial state f,g by

$$h = f + i\, A^{-1/2} g . \tag{10.10}$$

The representation is used in §13 below to determine the asymptotic behavior for $t \to \infty$ of the transient wave fields (10.7).

The R-B wave expansions of §8 can be used to construct the solutions wFE (10.7) and (10.9). To simplify the analysis it is assumed in the remainder of the work that G admits no R-B surface waves. In the general case most of the results derived below hold for states orthogonal to the subspace spanned by the surface waves. The scattering of R-B surface waves is not analyzed in this work.

Under the assumption of no surface waves the R-B wave expansions of the wave function (10.9) take the form

$$(10.11) \qquad v(t,X) = \ell.\text{i.m.} \int_{R_0^2} \psi_\pm(X,P)\, e^{-it\omega(P)}\, \hat{h}_\pm(P)\, dP$$

where the integral, together with its formal t-derivative ($= -i\, A^{1/2} v(t,X)$), converge in $L_2(G)$ (Theorems 8.5 and 8.6).

§11. Construction of the Wave Operators for A_p and $A_{0,p}$.

The purpose of the section is to prove the existence and completeness of the wave operators

$$(11.1) \qquad W_{\pm,p} = W_\pm(A_{0,p}^{1/2}, A_p^{1/2}, J_\Omega) = \text{s-}\lim_{t\to\pm\infty} e^{itA_{0,p}^{1/2}}\, J_\Omega\, e^{-itA_p^{1/2}}$$

where $J_\Omega : L_2(\Omega) \to L_2(B_0)$ is defined by

$$(11.2) \qquad J_\Omega\, h(X) = \begin{cases} h(X), & X \in \Omega, \\ 0, & X \in B_0 - \Omega. \end{cases}$$

This will be done by means of an explicit construction based on the eigenfunction expansions for A_p and $A_{0,p}$ of §6. The principal results are formulated as

<u>Theorem 11.1</u>. Let G be a grating domain of the class defined in §1. Let $p \in (-1/2, 1/2]$ and assume that $\sigma_0(A_p) = \phi$. Then $W_{+,p}$ and $W_{-,p}$ exist and are given by

$$(11.3) \qquad W_{\pm,p} = \Phi_{0,p}^*\, \Phi_{\mp,p}\ .$$

In particular, $W_{\pm,p} : L_2(\Omega) \to L_2(B_0)$ are unitary operators and one has

$$(11.4) \qquad \Pi_p(\lambda) = W_{\pm,p}^*\, \Pi_{0,p}(\lambda)\, W_{\pm,p} \quad \text{for all } \lambda \in R\ .$$

Theorem 11.1 is primarily of technical interest in the theory of scattering by diffraction gratings. It will be used in §12 to derive a construction of the wave operators for A and A_0.

Theorem 11.1 will be proved by the method of [30, Ch. 7]. Only the case of $W_{+,p}$ will be discussed, the other case being entirely similar. To begin consider the wave function

$$(11.5)\qquad v(t,\cdot) = e^{-itA_p^{1/2}} h \ , \quad h \in L_2(\Omega) \ .$$

The eigenfunction expansion theorem for A_p implies that $v(t,X)$ has the two representations

$$(11.6)\qquad v(t,X) = \underset{M\to\infty}{\ell.i.m.} \sum_{|m|\le M} \int_0^M \phi_\pm(X,p+m,q)\, e^{-it\omega(p+m,q)}\, \tilde{h}_\pm(p+m,q)\, dq \ ,$$

convergent in $L_2(\Omega)$. As in [30, Ch. 7], the incoming representation will be used to calculate the behavior of $v(t,\cdot)$ for $t \to +\infty$. The eigenfunction ϕ_- has the decomposition (see (6.4), (6.5))

$$(11.7)\qquad \phi_-(X,p+m,q) = j(y)\, \phi_0(X,p+m,q) + \phi_-'(X,p+m,q)$$

where ϕ_-' is incoming. Combining (11.6) and (11.7) gives

$$(11.8)\qquad v(t,X) = j(y)\, v_0^+(t,X) + v^+(t,X)$$

where

$$(11.9)\qquad v_0^+(t,X) = \underset{M\to\infty}{\ell.i.m.} \sum_{|m|\le M} \int_0^M \phi_0(X,p+m,q)\, e^{-it\omega(p+m,q)}\, \tilde{h}_-(p+m,q)\, dq$$

converges in $L_2(B_0)$ while

$$(11.10)\qquad v^+(t,X) = \underset{M\to\infty}{\ell.i.m.} \sum_{|m|\le M} \int_0^M \phi_-'(X,p+m,q)\, e^{-it\omega(p+m,q)}\, \tilde{h}_-(p+m,q)\, dq$$

converges in $L_2(\Omega)$. Note that the convergence of (11.6) and (11.9) implies that of (11.10). Moreover, $v_0^+(t,X)$ is a wave function in $L_2(B_0)$ for the reduced propagator $A_{0,p}$ of the degenerate grating; namely,

$$(11.11)\qquad v_0^+(t,\cdot) = e^{-itA_{0,p}^{1/2}}\, h_0^+$$

where $h_0^+ = v_0^+(0,\cdot) \in L_2(B_0)$ is given by

$$h_0^+ = \Phi_{0,p}^* \tilde{h}_- = \Phi_{0,p}^* \Phi_{-,p} h . \tag{11.12}$$

Theorem 11.1 will be shown to be a direct corollary of

Theorem 11.2. Under the hypotheses of Theorem 11.1 one has, for all $h \in L_2(\Omega)$,

$$\lim_{t\to+\infty} v^+(t,\cdot) = 0 \text{ in } L_2(\Omega) \tag{11.13}$$

and hence

$$\lim_{t\to+\infty} \|v(t,\cdot) - j(\cdot)\, v_0^+(t,\cdot)\|_{L_2(\Omega)} = 0 . \tag{11.14}$$

Proof of Theorem 11.2. Equation (11.14) can be written

$$\lim_{t\to+\infty} \left\| \left(e^{-itA_p^{1/2}} - J^* e^{-itA_{0,p}^{1/2}} \Phi_{0,p}^* \Phi_{-,p} \right) h \right\|_{L_2(\Omega)} = 0 \tag{11.15}$$

where $J^* : L_2(B_0) \to L_2(\Omega)$, the adjoint of the operator J defined by (6.22), is given by $J^* h(X) = j(y)\, h(X)|_\Omega$. Now the family of operators appearing in (11.15) is uniformly bounded for all $t \in R$. Hence to prove that (11.15) holds for all $h \in L_2(\Omega)$ it will suffice to verify it for all h from a dense subset of $L_2(\Omega)$. It will be convenient to use the dense subset $\mathcal{D}_0^- = \Phi_{-,p}^* \mathcal{D}_0$ where

$$\mathcal{D}_0 \subset \sum_{m\in Z} \oplus L_2(R_0) \tag{11.16}$$

is the set of all $g(q) = \{g_m(q) : m \in Z\}$ such that there is an $M = M(g)$ with the properties

$$g_m(q) \equiv 0 \text{ for } |m| > M , \text{ and} \tag{11.17}$$

$$g_m \in C_0^\infty(R_0 - E_{m,p}) \text{ for } |m| \le M \tag{11.18}$$

where $E_{m,p}$ is the exceptional set of (6.15). Moreover, it will suffice to verify (11.15) for functions of the form

$$h(X) = \int_0^\infty \phi_-(X,p+m,q)\, g(q)\, dq \tag{11.19}$$

where m is fixed and $g \in C_0^\infty(R_0 - E_{m,p})$ has support in an interval

$I \subset R_0 - E_{m,p}$, since the case of a general $h \in \mathcal{D}_0^-$ then follows by superposition. Thus the proof of Theorem 11.2 may be completed by showing that if

$$(11.20) \qquad v^+(t,X) = \int_I \phi'_-(X,p+m,q)\, e^{-it\omega(p+m,q)}\, g(q)\, dq$$

where $g \in C_0^\infty(R_0 - E_{m,p})$ and supp $g \subset I$ then (11.13) holds.

The definition of the function $\phi'_-(X,p+m,q)$, (6.4), (6.5), (6.13), (6.14), together with Theorem 4.15, implies that for fixed $m \in Z$ one has

$$(11.21) \qquad \phi'_- \in C(\Omega \times (R_0^2 - E))$$

where E is the exceptional set of (2.30). Moreover, the far-field form of ϕ'_- is

$$(11.22) \qquad \phi'_-(X,p+m,q) = \sum_{\ell \in L} a_\ell^-(p+m,q)\, e^{i(xp_\ell - yq_\ell)} + \rho_-(X,p+m,q)$$

where L is a finite set, independent of $q \in I$ (see (7.35)ff for the notation). Note that

$$(11.23) \qquad a_\ell^-(p+m,q)\, e^{-iyq_\ell} = \frac{1}{2\pi} \int_{-\pi}^{\pi} e^{-ix(p+\ell)}\, \phi'_-(x,y,p+m,q)\, dx .$$

It follows from (11.21) and (11.23) that

$$(11.24) \qquad a_\ell^-(\cdot+m,\cdot) \in C(R_0^2 - E) .$$

Moreover, by Lemma 7.3 there exists a constant $\mu = \mu(p,m) > 0$ and for each $r' > r > h$ a constant $C = C(I,p,m,r,r')$ such that, for p and m fixed,

$$(11.25) \qquad |\rho_-(X,p+m,q)| \le C\, e^{-\mu y} \text{ for all } X \in \Omega_{r'} \text{ and } q \in I .$$

Substitution of (11.22) in (11.20) gives

$$(11.26) \qquad v^+(t,X) = v_1^+(t,X) + v_2^+(t,X)$$

where

$$(11.27) \quad v_1^+(t,X) = \sum_{\ell \in L} \left\{ \int_I a_\ell^-(p+m,q)\, e^{-i(yq_\ell + t\omega(p+m,q))}\, g(q)\, dq \right\} e^{i(p+\ell)x} .$$

Note that, by (11.24), each $a_\ell^-(p+m,\cdot)$ is continuous on the closed interval

I. Now each of the integrals in (11.27) has the form of a modal wave in a simple waveguide; cf. (7.63)ff. It follows from (7.69) applied to the finite sum in (11.27) that

$$\lim_{t\to+\infty} \|v_1^+(t,\cdot)\|_{L_2(B_0)} = 0 . \tag{11.28}$$

It remains to show that $v_2^+(t,\cdot) \to 0$ in $L_2(\Omega)$ when $t \to +\infty$. This will be done by applying Lemma 7.5 to $u = v_2^+ = v^+ - v_1^+$. To this end note that for all $t \in R$ one has $v^+(t,\cdot) \in L_2(\Omega)$ by (11.10) and $v_1^+(t,\cdot) \in L_2(B_0)$. Thus $v_+^2(t,\cdot) = v^+(t,\cdot) - v_1^+(t,\cdot) \in L_2(\Omega)$ for all $t \in R$ which verifies (7.75).

The local decay property (7.76) follows from the local compactness property of the grating domain G, assumed in §1, and the abstract decay theorem of [29]. The proof is the same as that of [30, Theorem 5.5] and is therefore not repeated here.

Finally, (7.77) follows directly from the estimate (11.25) and the representation

$$v_2^+(t,X) = \int_I \rho_-(X,p+m,q)\, e^{-it\omega(p+m,q)}\, g(q)\, dq \tag{11.29}$$

which imply

$$|v_2^+(t,X)| \le C\, e^{-\mu y} \int_I |g(q)|\, dq \text{ for all } X \in \Omega_r, \text{ and } t \in R . \tag{11.30}$$

This completes the proof of Theorem 11.2.

<u>Proof of Theorem 11.1</u>. The proof follows that of [30, Corollary 7.2]. In fact, the calculation given there, adapted to the present problem, gives the estimate

$$\begin{aligned} &\left\| \left(J_\Omega\, e^{-itA_p^{1/2}} - e^{-itA_{0,p}^{1/2}}\, \Phi_{0,p}^* \Phi_{-,p} \right) h \right\|_{L_2(B_0)} \\ &\le \left\| \left(e^{-itA_p^{1/2}} - J_\Omega^*\, e^{-itA_{0,p}^{1/2}}\, \Phi_{0,p}^* \Phi_{-,p} \right) h \right\|_{L_2(\Omega)} \\ &\quad + \left\| e^{-itA_{0,p}^{1/2}} (\Phi_{0,p}^* \Phi_{-,p} h) \right\|_{L_2(B_{0,r})} + \left\| e^{-itA_p^{1/2}} h \right\|_{L_2(\Omega_{0,r})} . \end{aligned} \tag{11.31}$$

The first term on the right in (11.31) tends to zero when $t \to +\infty$ by (11.15). The last two terms tend to zero when $t \to +\infty$ by the local decay property used in the proof of Theorem 11.2. It follows that the left hand side of (11.31) tends to zero when $t \to +\infty$ which proves the existence of $W_{\pm,p}$ and equation (11.3). Finally, to verify (11.4) note that it can be written

$$\Pi_p(\lambda) = \Phi_{\pm,p}^* \, \Phi_{0,p} \, \Pi_{0,p}(\lambda) \, \Phi_{0,p}^* \, \Phi_{\pm,p} \quad \text{for } \lambda \in R \tag{11.32}$$

by (11.3). The unitarity of $\Phi_{\pm,p}$ implies that an equivalent relation is

$$\Phi_{\pm,p} \, \Pi_p(\lambda) \, \Phi_{\pm,p}^* = \Phi_{0,p} \, \Pi_{0,p}(\lambda) \, \Phi_{0,p}^* \quad \text{for } \lambda \in R \,. \tag{11.33}$$

But this last equation is correct because the two sides coincide with the operation

$$\{g_m(q)\} \to \{H(\lambda - \omega^2(p+m,q)) \; g_m(q)\} \tag{11.34}$$

in $\Sigma \oplus L_2(R_0)$; see Theorem 6.7. This completes the proof.

§12. Construction of the Wave Operators for A and A_0.

The purpose of the section is to prove the existence and completeness of the wave operators

$$W_\pm = W_\pm(A_0^{1/2}, A^{1/2}, J_G) = \operatorname*{s-lim}_{t\to\pm\infty} e^{itA_0^{1/2}} \, J_G \, e^{-itA^{1/2}} \tag{12.1}$$

where $J_G : L_2(G) \to L_2(R_0^2)$ is defined by

$$J_G \, h(X) = \begin{cases} h(X) \,, & X \in G \,, \\ 0 \,, & X \in R_0^2 - G \,. \end{cases} \tag{12.2}$$

The principal results of the section are formulated as

Theorem 12.1. Let G be a grating domain of the class defined in §1 and let $A = A(G)$ admit no surface waves. Then W_+ and W_- exist and are given by

$$W_\pm = \Phi_0^* \, \Phi \quad . \tag{12.3}$$

In particular, $W_\pm : L_2(G) \to L_2(R_0^2)$ are unitary operators and one has

$$\Pi(\lambda) = W_\pm^* \, \Pi_0(\lambda) \, W_\pm \quad \text{for all } \lambda \in R \,. \tag{12.4}$$

The proof of Theorem 12.1 will be based on Theorem 11.1 and a series of lemmas that relate the grating propagators A and A_0 to the corresponding families of reduced propagators A_p and $A_{0,p}$, $-1/2 < p \leq 1/2$.

The Mapping U. As a first step, the correspondence introduced in (8.10) will be extended to a unitary mapping $U : L_2(G) \to L_2((-1/2,1/2],L_2(\Omega))$. To see how this may be done note that if $f \in L_2(G)$ then

$$f(x+2\pi\ell,y)|_{\Omega} \in L_2(\Omega) \text{ for all } \ell \in Z, \text{ and} \tag{12.5}$$

$$\sum_{\ell \in Z} \|f(\cdot+2\pi\ell,\cdot)\|^2_{L_2(\Omega)} = \|f\|^2_{L_2(G)} < \infty . \tag{12.6}$$

Hence the Plancherel theory in the Lebesgue space $L_2((-1/2,1/2],L_2(\Omega))$ implies that the Fourier series

$$F(x,y,p) = \sum_{\ell \in Z} e^{-2\pi i\ell p} f(x+2\pi\ell,y)|_{\Omega} \tag{12.7}$$

converges in this space and Parseval's relation is valid. Combining this result and (12.6) gives

$$\|F\|_{L_2((-1/2,1/2],L_2(\Omega))} = \|f\|_{L_2(G)} \tag{12.8}$$

for all $f \in L_2(G)$.

Lemma 12.2. The mapping $U : L_2(G) \to L_2((-1/2,1/2],L_2(\Omega))$ defined by $Uf = F$ and (12.7) is unitary.

Proof. The preceding discussion implies that Uf is defined for all $f \in L_2(G)$ and U is isometric. The surjectivity of U follows from the Plancherel theory. Indeed, every $F \in L_2((-1/2,1/2],L_2(\Omega))$ has a Fourier development

$$F(X,p) = \sum_{\ell \in Z} e^{-2\pi i\ell p} F_{\ell}(X) , \tag{12.9}$$

convergent in $L_2((-1/2,1/2],L_2(\Omega))$. The Fourier coefficients in (12.9) are defined by the Bochner integrals

$$F_{\ell}(X) = \int_{-1/2}^{1/2} e^{2\pi i\ell p} F(X,p)\, dp \in L_2(\Omega) \tag{12.10}$$

and Parseval's relation holds:

$$\|F\|^2_{L_2((-1/2,1/2],L_2(\Omega))} = \sum_{\ell\in Z} \|F_\ell\|^2_{L_2(\Omega)} \; . \tag{12.11}$$

Thus to construct $f = U^{-1}F$ one need only require that $f(x+2\pi\ell,y)|_\Omega = F_\ell(X) \in L_2(\Omega)$, or

$$f(X)|_{\Omega(\ell)} = F_\ell(x-2\pi\ell,y) \text{ for all } \ell \in Z \; . \tag{12.12}$$

Parseval's relation then guarantees that $f \in L_2(G)$ and (12.12) implies that $Uf = F$.

The next lemma makes it possible to construct operators $\Psi(A)$ from the corresponding reduced operators $\Psi(A_p)$ with $p \in (-1/2,1/2]$.

<u>Lemma 12.3</u>. For all bounded Borel functions $\Psi(\lambda)$ defined for $\lambda \geq 0$ and for all $f \in L_2(G)$ one has

$$(U\,\Psi(A)f)(\cdot,p) = \Psi(A_p)\; Uf(\cdot,p) \in L_2(\Omega) \tag{12.13}$$

for almost every $p \in (-1/2,1/2]$.

<u>Proof</u>. The result will be derived from the R-B wave expansions for A and A_p and the corresponding Plancherel relations. To this end let $f,g \in L_2(G)$ and write $Uf = F$ and $Ug = G$. Moreover, assume that $\Phi_+ g = \hat{g}_+ \in L_2^{com}(R_0^2)$. Then Lemma 12.2 and the results of §6 and §8 imply

$$\begin{aligned}
(U\,\Psi(A)f,G) &= (\Psi(A)f,g) = (\Phi_+\,\Psi(A)f,\; \Phi_+ g) = \int_{R_0^2} \overline{\Psi(\omega^2(P))\; \hat{f}_+(P)}\; \hat{g}_+(P)\, dP \\
&= \sum_{m\in Z} \int_{m-1/2}^{m+1/2} \int_0^\infty \overline{\Psi(\omega^2(p,q))\; \hat{f}_+(p,q)}\; \hat{g}_+(p,q)\, dq dp \\
&= \sum_{m\in Z} \int_{-1/2}^{1/2} \int_0^\infty \overline{\Psi(\omega^2(p+m,q))\; \hat{f}_+(p+m,q)}\; \hat{g}_+(p+m,q)\, dq dp \\
&= \int_{-1/2}^{1/2} \left\{ \sum_{m\in Z} \int_0^\infty \overline{\Psi(\omega^2(p+m,q))\; \tilde{F}_+(p+m,q,p)}\; \tilde{G}_+(p+m,q,p) dq \right\} dp \\
&= \int_{-1/2}^{1/2} (\Psi(A_p)\; F(\cdot,p),G(\cdot,p))_{L_2(\Omega)}\; dp \; .
\end{aligned} \tag{12.14}$$

Note that the hypothesis $\hat{g}_+ \in L_2^{com}(R_0^2)$ implies that the m-summation and interval of q-integration in (12.14) are finite. Moreover, since such functions $\hat{g}_+$ are dense in $L_2(R_0^2)$ the relation (12.14) holds for all

$G \in L_2((-1/2,1/2],L_2(\Omega))$. On taking $G(X,p) = G_1(X)\, G_2(p)$ in (12.14) where $G_1 \in L_2(\Omega)$ and $G_2(p) \in L_2(-1/2,1/2]$ are arbitrary one gets (12.13).

The mapping U obviously depends on the grating domain $G : U = U_G$. In the special case $G = R_0^2$ let $U_0 = U_{R_0^2}$. With this notation one has

Lemma 12.4. The operators U, U_0, J_G and J_Ω satisfy

$$(12.15) \qquad U_0\, J_G = J_\Omega\, U \,.$$

Proof. The definition of U implies that

$$(12.16) \qquad (J_\Omega\, Uf)(X,p) = \sum_{\ell \in Z} e^{-2\pi i \ell p}\, J_\Omega\, (f(x+2\pi\ell,y)|_\Omega) \,,$$

$$(12.17) \qquad (U_0\, J_G\, f)(X,p) = \sum_{\ell \in Z} e^{-2\pi i \ell p}\, (J_G\, f(x+2\pi\ell,y)|_{B_0}) \,,$$

for all $f \in L_2(G)$. These obviously define the same function, which implies (12.15).

The next lemma will be used to relate the wave operators for A and A_0 to those for A_p and $A_{0,p}$.

Lemma 12.5. For all $f \in L_2(G)$ one has

$$(12.18) \qquad U_0(\Phi_0^* \Phi_\pm f)(\cdot,p) = \Phi_{0,p}^* \Phi_{\pm,p}\, Uf(\cdot,p) \in L_2(B_0)$$

for almost every $p \in (-1/2,1/2]$.

Proof. The relation (8.11) can be written

$$(12.19) \qquad (\Phi_\pm f)(p+m,q) = (\Phi_{\pm,p}\, F(\cdot,p))_m(q) = (\Phi_{\pm,p}\, Uf(\cdot,p))_m(q) \,.$$

The relation was proved for all $f \in L_2^{com}(G)$. However (12.19), as a relationship in $\Sigma \oplus L_2(R_0)$, extends immediately from the dense set $L_2^{com}(G)$ to all of $L_2(G)$. In particular, specializing (12.19) to $G = R_0^2$ gives

$$(12.20) \qquad (\Phi_0 f_0)(p+m,q) = (\Phi_{0,p}\, U_0\, f_0(\cdot,p))_m(q)$$

for all $f_0 \in L_2(R_0^2)$. Substituting $f_0 = \Phi_0^* \Phi_\pm f$ in (12.20) gives

$$(12.21) \quad (\Phi_{0,p}(U_0\, \Phi_0^* \Phi_\pm f)(\cdot,p))_m(q) = \Phi_\pm f(p+m,q) = (\Phi_{\pm,p}\, Uf(\cdot,p))_m(q)$$

in $\Sigma \oplus L_2(R_0)$, by (12.19). Thus

$$\Phi_{0,p}(U_0\, \Phi_0^* \, \Phi_\pm f)(\cdot,p) = \Phi_{\pm,p}\, Uf(\cdot,p) \tag{12.22}$$

for almost every $p \in (-1/2,1/2]$. (12.22) is equivalent to (12.18).

Lemmas 12.2-12.5 will now be shown to imply Theorem 12.1. The main step in the proof is described by

Theorem 12.6. For every $h \in L_2(G)$ and $H = Uh$ one has

$$\begin{aligned} &\left\| J_G\, e^{-itA^{1/2}} h - e^{-itA_0^{1/2}} \Phi_0^* \, \Phi_\pm h \right\|^2_{L_2(R_0^2)} \\ &= \int_{-1/2}^{1/2} \left\| J_\Omega\, e^{-itA_p^{1/2}} H(\cdot,p) - e^{-itA_{0,p}^{1/2}} \Phi_{0,p}^* \, \Phi_{\pm,p}\, H(\cdot,p) \right\|^2_{L_2(B_0)} dp\,. \end{aligned} \tag{12.23}$$

Proof. Lemma 12.2 implies that

$$\begin{aligned} &\left\| J_G\, e^{-itA^{1/2}} h - e^{-itA_0^{1/2}} \Phi_0^* \, \Phi_\pm h \right\|^2_{L_2(R_0^2)} \\ &= \int_{-1/2}^{1/2} \left\| U_0 (J_G\, e^{-itA^{1/2}} h - e^{-itA_0^{1/2}} \Phi_0 \, \Phi_\pm h)(\cdot,p) \right\|^2_{L_2(B_0)} dp\,. \end{aligned} \tag{12.24}$$

Moreover, Lemmas 12.4 and 12.3 imply

$$(U_0\, J_G\, e^{-itA^{1/2}} h)(\cdot,p) = (J_\Omega\, U\, e^{-itA^{1/2}} h)(\cdot,p) = J_\Omega\, e^{-itA_p^{1/2}} U h(\cdot,p)\,. \tag{12.25}$$

Finally, Lemmas 12.3 and 12.5 imply

$$\begin{aligned} (U_0\, e^{-itA_0^{1/2}} \Phi_0^* \, \Phi_\pm h)(\cdot,p) &= e^{-itA_{0,p}^{1/2}} U_0 (\Phi_0^* \, \Phi_\pm h)(\cdot,p) \\ &= e^{-itA_{0,p}^{1/2}} \Phi_{0,p}^* \, \Phi_{\pm,p}\, U h(\cdot,p)\,. \end{aligned} \tag{12.26}$$

Combining (12.24), (12.25) and (12.26) gives (12.23).

Proof of Theorem 12.1. Lemma 12.2 implies that $H(\cdot,p) \in L_2(\Omega)$ for almost every $p \in (-1/2,1/2]$. Hence the integrand on the right hand side of (12.23) tends to zero when $t \to \mp\infty$ by Theorem 11.1 (see (11.31)). Moreover, the operators appearing in the integrand are all bounded with bound 1 and hence one has

$$\left\| J_\Omega \, e^{-itA_p^{1/2}} H(\cdot,p) - e^{-itA_{0,p}^{1/2}} \Phi_{0,p}^{*} \, \Phi_{\pm,p} H(\cdot,p) \right\|_{L_2(B_0)} \le 2 \, \|H(\cdot,p)\|_{L_2(\Omega)} \tag{12.27}$$

for all $t \in R$ and almost every $p \in (-1/2,1/2]$. Thus the existence of W_+ and W_-, and the relation (12.3), follow from (12.23) and Lebesgue's dominated convergence theorem. The final statement of Theorem 12.1, equation (12.4), follows from (12.3) and the eigenfunction expansions for A and A_0, exactly as in the proof of Theorem 11.1.

§13. Asymptotic Wave Functions and Energy Distributions

In this section the existence of the wave operators $W_\pm$ is shown to imply that transient wave fields in grating domains G are asymptotically equal in the energy norm, for $t \to +\infty$, to transient wave fields in the degenerate grating domain R_0^2. The latter are then shown to be the restrictions to R_0^2 of free waves in R^2. Such free waves possess asymptotic wave functions in the sense of the author's monograph on scattering by bounded obstacles [30]. These results are shown below to imply that transient wave fields $u(t,X)$ with finite energy in grating domains possess asymptotic wave functions

$$u_k^\infty(t,X) = r^{-1/2} F_k(r-t,\theta) \;, \quad k = 0,1,2 \;, \tag{13.1}$$

(where $X = (r\cos\theta, r\sin\theta)$) such that if $(t,x,y) = (X_0,X_1,X_2)$ and $D_k = \partial/\partial X_k$ for $k = 0,1,2$ then

$$\lim_{t\to\infty} \|D_k u(t,\cdot) - u_k^\infty(t,\cdot)\|_{L_2(G)} = 0 \;, \quad k = 0,1,2 \;. \tag{13.2}$$

Moreover, the waveforms $F_k(\tau,\theta)$ are calculated from the initial state $f(X)$, $g(X)$ of $u(t,X)$. Finally, (13.2) and the results of [30, Ch. 8] are used to calculate the asymptotic distributions of energy for transient wave fields in grating domains.

The starting point for the calculation of the asymptotic wave functions (13.1) is the complex-valued wave function $v(t,X)$ defined by (10.9), (10.10). The existence of the wave operator W_+ defined by (12.1) implies that

$$e^{-itA^{1/2}} h \sim e^{-itA_0^{1/2}} W_+ h \;, \quad t \to +\infty \;, \tag{13.3}$$

in the sense of convergence in $L_2(G)$. Moreover, if $h \in D(A^{1/2})$ then the analogue of (13.3) holds for the first derivatives. This result may be formulated as a generalization of the corresponding result for exterior domains [30, Theorem 7.5], as follows.

Theorem 13.1. Let G satisfy the hypotheses of Theorem 12.1 and let $h \in D(A^{1/2})$. Then $v(t,\cdot) = e^{-itA^{1/2}} h$ is a solution wFE in G, $h_{sc} = W_+ h \in D(A_0^{1/2})$, $v_0(t,\cdot) = e^{-itA_0^{1/2}} h_{sc}$ is a solution wFE in R_0^2 and

$$(13.4) \qquad \lim_{t\to+\infty} \|D_k\, v(t,\cdot) - D_k\, v_0(t,\cdot)\|_{L_2(G)} = 0 \text{ for } k = 0,1,2 \,.$$

The proof is precisely the same as the one for exterior domains given in [30] and is therefore omitted.

The initial state $h_{sc} = W_+ h$ for the wave field $v_0(t,X)$ satisfies

$$(13.5) \qquad (h_{sc})\hat{}_0 = \Phi_0\, h^+ = \Phi_- h = \hat{h}_-$$

by (12.3). Thus v_0 has the R-B wave representation ((10.11) for A_0)

$$(13.6) \qquad v_0(t,X) = \int_{R_0^2} \psi_0(X,P)\, e^{-it\omega(P)}\, \hat{h}_-(P)\, dP \,.$$

To show that $v_0(t,X)$ has a continuation to a wave field wFE in R^2 the Neumann and Dirichlet cases will be treated separately.

The Neumann Case. Here one has (see (2.31) for normalization)

$$(13.7) \qquad \psi_0(X,P) = \psi_{0\pm}^N(X,P) = \psi_0^N(X,P) = \frac{1}{2\pi}\, e^{ipx}(e^{iqy} + e^{-iqy})$$

and substitution in (13.6) gives, after a simple transformation,

$$(13.8) \qquad v_0(t,X) = \frac{1}{2\pi} \int_{R^2} e^{i(xp+yq-t\omega(p,q))}\, \hat{h}_0(p,q)\, dpdq$$

where

$$(13.9) \qquad \hat{h}_0(p,q) = \begin{cases} \hat{h}_-(p,q)\,, & (p,q) \in R_0^2\,, \\ \hat{h}_-(p,-q)\,, & (p,-q) \in R_0^2\,. \end{cases}$$

The Dirichlet Case. Here if ψ_0 is normalized by (see (2.31))

$$(13.10) \qquad \psi_0(X,P) = \psi_{0-}^D(X,P) = i\,\psi_0^D(X,P) = \frac{1}{2\pi}\, e^{ipx}(e^{iqy} - e^{-iqy})$$

then substitution in (13.6) again gives (13.8), but with

$$(13.11) \qquad \hat{h}_0(p,q) = \begin{cases} \hat{h}\_(p,q) , & (p,q) \in R_0^2 , \\ -\hat{h}\_(p,-q) , & (p,-q) \in R_0^2 . \end{cases}$$

Thus in both cases $v_0(t,X)$ has a continuation (13.8) to a wave field in R^2. Moreover, the hypothesis $h \in D(A^{1/2})$ of Theorem 13.1 implies that $\hat{h}\_(p,q)$ and $\sqrt{p^2+q^2}\,\hat{h}\_(p,q)$ are in $L_2(R_0^2)$ and hence $\hat{h}_0(p,q)$ and $\sqrt{p^2+q^2}\,\hat{h}_0(p,q)$ are in $L_2(R^2)$. It follows that the extended wave field (13.8) is a solution wFE in R^2. Thus the results of [30, Ch. 2] are applicable and allow the construction of asymptotic wave functions

$$(13.12) \qquad v_k^\infty(t,X) = r^{-1/2} H_k(r-t,\theta) , \quad k = 0,1,2 ,$$

such that

$$(13.13) \qquad \lim_{t\to+\infty} \|D_k v_0(t,\cdot) - v_k^\infty(t,\cdot)\|_{L_2(R^2)} = 0 , \quad k = 0,1,2 .$$

By restricting the functions to G one obtains

<u>Corollary 13.2</u>. Under the hypotheses of Theorem 13.1 one has

$$(13.14) \qquad \lim_{t\to+\infty} \|D_k v(t,\cdot) - v_k^\infty(t,\cdot)\|_{L_2(G)} = 0 , \quad k = 0,1,2 ,$$

where the function v_k^∞ are given by (13.12) with waveforms H_k defined by

$$(13.15) \quad H_0(\tau,\theta) = \frac{\text{l.i.m.}}{(2\pi)^{1/2}} \int_0^\infty e^{i\tau\omega}\, \hat{h}\_(\omega\cos\theta, \omega\sin\theta)(-i\omega)^{1/2}\, d\omega ,$$

convergent in $L_2(R \times [0,\pi])$, and

$$(13.16) \qquad H_1(\tau,\theta) = -H_0(\tau,\theta)\cos\theta , \quad H_2(\tau,\theta) = -H_0(\tau,\theta)\sin\theta .$$

Equations (13.14) follow from (13.4) and (13.13) by the triangle inequality. Equations (13.13), (13.15) and (13.16) follow directly from [3, Ch. 2]; see the proof of [3, Theorem 2.10].

To obtain corresponding results for the real-valued wave field $u(t,X)$ generated by the initial state f,g one need only take the real part of $v(t,X)$ and use equation (10.10) which relates h to f and g. This leads to

<u>Theorem 13.3</u>. Let G satisfy the hypotheses of Theorem 12.1. Let $f \in D(A^{1/2})$ and $g \in L_2(G)$ and define asymptotic wave functions

(13.17) $$u_k^\infty(t,X) = r^{-1/2} F_k(r-t,\theta) , \quad k = 0,1,2 ,$$

by

$$F_0(\tau,\theta) = \text{Re} \left\{\frac{1}{(2\pi)^{1/2}} \int_0^\infty e^{i\tau\omega}[\hat{g}\_(\omega \cos \theta,\omega \sin \theta)-i\omega\hat{f}\_(\omega \cos \theta,\omega \sin \theta)](-i\omega)^{1/2} \, d\omega\right\}$$

(13.18)

convergent in $L_2(R \times [0,\pi])$, and

(13.19) $$F_1(\tau,\theta) = -F_0(\tau,\theta) \cos \theta , \quad F_2(\tau,\theta) = -F_0(\tau,\theta) \sin \theta .$$

Then the solution wFE (10.7) generated by f and g satisfies

(13.20) $$\lim_{t\to+\infty} \|D_k u(t,\cdot) - u_k^\infty(t,\cdot)\|_{L_2(G)} = 0 , \quad k = 0,1,2 .$$

Proof. To begin, assume that $g \in D(A^{-1/2})$ and define h by (10.10). Then $h \in D(A^{1/2})$ and Corollary 13.2 is applicable to $v(t,\cdot) = e^{-itA^{1/2}} h$. Moreover,

(13.21) $$|P| \hat{h}\_(P) = |P| \hat{f}\_(P) + i \hat{g}\_(P)$$

which implies that $F_k = \text{Re } H_k$ and hence

(13.22) $$u_k^\infty(t,X) = \text{Re } \{v_k^\infty(t,X)\} , \quad k = 0,1,2 .$$

Thus (13.20) follows from (13.14) by the triangle inequality. To remove the restriction that $g \in D(A^{-1/2})$ note that $D(A^{-1/2})$ is dense in $L_2(G)$, by the spectral theorem. Moreover, one has (see [30, Theorem 2.5])

(13.23)
$$\begin{aligned} \|u_0^\infty(t,\cdot)\|_{L_2(G)} &\le \|v_0^\infty(t,\cdot)\|_{L_2(R_0^2)} \le \|H_0\| = \|\hat{H}_0\| \\ &= \left(\int_0^\infty \int_0^\pi \omega^3 \, |\hat{h}\_(\omega \cos \theta,\omega \sin \theta)|^2 \, d\theta \, d\omega\right)^{1/2} \\ &= \left(\int_{R_0^2} |P|^2 \, |\hat{h}\_(P)|^2 \, dP\right)^{1/2} = \| |P| \, \hat{h}\_(P)\|_{L_2(R_0^2)} \\ &\le \| |P| \, \hat{f}\_(P)\| + \|\hat{g}\_\| = \|A^{1/2} f\| + \|g\| . \end{aligned}$$

Similarly,

$$\|u_k^\infty(t,\cdot)\|_{L_2(G)} \le \|u_0^\infty(t,\cdot)\| \le \|A^{1/2} f\| + \|g\| \;, \quad k = 1,2 \;. \tag{13.24}$$

Finally, the conservation of energy theorem implies that

$$\begin{aligned} \|D_k u(t,\cdot)\|_{L_2(G)} &\le E(u,G,t)^{1/2} = E(u,G,0)^{1/2} \\ &= (\|A^{1/2} f\|^2 + \|g\|^2)^{1/2} \le \|A^{1/2} f\| + \|g\| \;, \\ & \qquad k = 0,1,2 \;. \end{aligned} \tag{13.25}$$

It follows from (13.23), (13.24) and (13.25) that (13.20) can be extended to all $f \in D(A^{1/2})$ and $g \in L_2(G)$ by a well-known density argument (see, e.g., [30, Proof of Theorem 2.6]).

Theorem 13.3 permits the extension to grating domains of the results on asymptotic energy distributions in exterior domains given in [3]. The principal results are formulated below. The proofs are identical to those of [30] and are therefore omitted.

Corollary 13.4 (Scattering into Cones). Let

$$\Gamma = \{X = (r \cos \theta, r \sin \theta) : r > 0 \text{ and } \theta \in \Gamma_0\} \tag{13.26}$$

where Γ_0 is a Lebesgue-measurable subset of $[0,\pi]$, and let $X_0 \in R^2$. Then under the hypotheses of Theorem 13.3 the limit

$$E^\infty(u,G \cap (\Gamma + X_0)) = \lim_{t\to+\infty} E(u,G \cap (\Gamma + X_0),t) \tag{13.27}$$

exists and

$$E^\infty(u,G \cap (\Gamma + X_0)) = \int_\Gamma ||P|\ \hat{f}_-(P) + i\ \hat{g}_-(P)|^2\ dP \;. \tag{13.28}$$

Corollary 13.5 (Transiency of Energy in Slabs). Let

$$\Sigma = \{X : d_1 \le X \cdot X_0 \le d_2\} \tag{13.29}$$

where d_1 and d_2 $(> d_1)$ are constants and $X_0 \in R^2$ is a unit vector. Then under the hypotheses of Theorem 13.3 one has

(13.30) $$\lim_{t\to+\infty} E(u, G\cap\Sigma, t) = 0 .$$

Note that Corollary 13.5 implies the transiency of the energy in bounded sets since every bounded set $K \subset R^2$ is contained in a slab (13.29).

§14. Construction and Structure of the S-Matrix

The scattering operator associated with the pair A, A_0 is the linear operator $S : L_2(R_0^2) \to L_2(R_0^2)$ defined by

(14.1) $$S = W_+ W_-^* .$$

The corresponding operator in $L_2(R_0^2)$ defined by

(14.2) $$\hat{S} = \Phi_0 S \Phi_0^*$$

is the Heisenberg operator, or S-matrix, for the pair A, A_0. From the representation $W_\pm = \Phi_0^* \Phi_\mp$ of Theorem 12.1 one has

(14.3) $$\hat{S} = \Phi_- \Phi_+^* .$$

The unitarity of $W_\pm$ and Φ_0 imply that S and $\hat{S}$ are unitary operators in $L_2(R_0^2)$. The purpose of this section is to calculate $\hat{S}$. Specifically, it will be shown how $\hat{S}$ can be constructed from the scattering coefficients $\{c_\ell^\pm(p,q)\}$ of the R-B waves $\psi_\pm(x,y,p,q)$ and the relationships among these coefficients imposed by the unitarity of S will be determined. The role of the S-matrix in the scattering of transient fields by gratings will be developed in §15.

If $h \in L_2(G)$ then (14.3) implies that the functions $\hat{h}_+ = \Phi_+ h$ and $\hat{h}_- = \Phi_- h$ satisfy

(14.4) $$\hat{h}_- = \hat{S}\, \hat{h}_+ .$$

Thus $\hat{S}$ may be calculated by calculating the relationship between $\hat{h}_-$ and $\hat{h}_+$. This will be done by using the incoming and outgoing R-B wave representations of $v(t,\cdot) = e^{-itA^{1/2}} h$ to calculate in two different ways the asymptotic wave function in $L_2(G)$ associated with $v(t,\cdot)$; say

(14.5) $$v^\infty(t,X) = r^{-1/2} H(r-t,\theta) .$$

The function $H \in L_2(R \times [0,\pi])$ is uniquely determined by the condition

(14.6) $$\lim_{t\to+\infty} \|v(t,\cdot) - v^{\infty}(t,\cdot)\|_{L_2(G)} = 0 \ ;$$

see [30, Theorem 2.5]. The equality of the representations of H obtained from the incoming and outgoing representations of $v(t,\cdot)$ provides the required relationship between $\hat{h}_-$ and $\hat{h}_+$.

<u>First Calculation of H</u>. Theorem 12.1 implies that

(14.7) $$\lim_{t\to+\infty} \|v(t,\cdot) - v_0(t,\cdot)\|_{L_2(G)} = 0$$

where $v_0(t,\cdot) = e^{-itA_0^{1/2}} h_{sc}$ is the wave function in $L_2(R_0^2)$ of Theorem 13.1. Proceeding as in the proof of Corollary 13.2 one shows that (14.5), (14.6) hold with

(14.8) $$H(\tau,\theta) = \frac{1}{(2\pi)^{1/2}} \int_0^{\infty} e^{i\tau\omega}\, \hat{h}_-(\omega\cos\theta,\omega\sin\theta)(-i\omega)^{1/2}\, d\omega \ .$$

The convergence $v_0(t,\cdot) - v^{\infty}(t,\cdot) \to 0$ in $L_2(R^2)$ was proved in [30, Theorem 2.6].

<u>A Classification of the R-B Waves</u>. The second calculation of H will be based on the outgoing representation

(14.9) $$v(t,X) = \ell.i.m. \int_{R_0^2} \psi_+(X,P)\, e^{-it\omega(P)}\, \hat{h}_+(P)\, dP \ .$$

The R-B wave ψ_+ has the expansion for $y \geq h$, (2.26)

(14.10)
$$\psi_+(x,y,p,q) = (2\pi)^{-1} e^{i(px-qy)} + (2\pi)^{-1} \sum_{(p+\ell)^2<p^2+q^2} c_\ell^+(p,q)\, e^{i(p_\ell x+q_\ell y)}$$
$$+ (2\pi)^{-1} \sum_{(p+\ell)^2\geq p^2+q^2} c_\ell^+(p,q)\, e^{ip_\ell x}\, e^{-\{(p+\ell)^2-p^2-q^2\}^{1/2} y}$$

where

(14.11) $$(p_\ell,q_\ell) = (p+\ell,\{p^2+q^2-(p+\ell)^2\}^{1/2}) \ .$$

The first sum in (14.10) is a superposition of a finite number of outgoing plane waves, while the second sum is an exponentially decreasing function of y for $(x,y) \in R_0^2 - E$ (Lemma 7.3). In the calculation of the asymptotic wave function (14.5) from (14.9) and (14.10) a difficulty arises because the number of terms in the first sum varies with $(p,q) \in R_0^2$. This number changes at the points $(p,q) \in E$ and is constant on the components of the set $R_0^2 - E$. It will therefore be convenient to classify the R-B waves by means of these components. Note that $(p,q) \in E$ if and only if $q > 0$ and

$$q_\ell^2 \equiv p^2 + q^2 - (p + \ell)^2 = 0 \ , \ \ell \in Z - \{0\} \ . \tag{14.12}$$

The set π_ℓ so defined is the portion lying in R_0^2 of the parabola with focus at (0,0) and vertex at $(-\ell/2,0)$. The curves π_ℓ and π_m are disjoint if $\ell m > 0$ and intersect orthogonally if $\ell m < 0$. Thus if

$$\begin{aligned} \mathcal{O}_m &= R_0^2 \cap \{(p,q) : |p + m| < \sqrt{p^2 + q^2} < |p + m + 1|\} \ , \\ \mathcal{O}_{-n} &= R_0^2 \cap \{(p,q) : |p - n| < \sqrt{p^2 + q^2} < |p - n - 1|\} \ , \end{aligned} \tag{14.13}$$

where $m,n = 0,1,2,\cdots$, then $\mathcal{O}_m$ is the domain between π_m and π_{m+1}, $\mathcal{O}_{-n}$ is the domain between π_{-n} and π_{-n-1} and the sets

$$\mathcal{O}_{m,n} = \mathcal{O}_m \cap \mathcal{O}_{-n} \ , \quad m,n = 0,1,2,\cdots \ , \tag{14.14}$$

are the components of $R_0^2 - E$:

$$R_0^2 - E = \bigcup_{m,n=0}^{\infty} \mathcal{O}_{m,n} \ . \tag{14.15}$$

Note that $(p,q) \in \mathcal{O}_{m,n}$ if and only if the expansion (14.10) of $\psi_+(x,y,p,q)$ contains exactly $m + n + 1$ outgoing plane waves with the propagation directions (p_ℓ,q_ℓ), $-n \le \ell \le m$. (Note $\mathcal{O}_{-0} \ne \mathcal{O}_0$, $\mathcal{O}_0 \cap \mathcal{O}_{-0} = \mathcal{O}_{0,0}$.)

<u>Second Calculation of H</u>. In calculating $\hat{S}$ it will suffice to determine $\hat{S}\,\hat{h}_+$ for functions $\hat{h}_+$ of a dense set in $L_2(R_0^2)$ because $\hat{S}$ is known to be unitary. For this purpose it will be convenient to use functions $\hat{h}_+ \in C_0^\infty(R_0^2 - E)$. For such functions, supp $\hat{h}_+$ is a compact subset of the set (14.15). Hence, supp $\hat{h}_+$ meets only finitely many of the sets $\mathcal{O}_{m,n}$ and each component of supp $\hat{h}_+$ lies in one of these sets. Thus in calculating $\hat{S}\,\hat{h}_+$ it will be enough to consider the case where

$$\text{supp } \hat{h}_+ = K \subset \mathcal{O}_{m,n} \ , \quad m \text{ and } n \text{ fixed} \ . \tag{14.16}$$

The case of a general $\hat{h}_+ \in C_0^\infty(R_0^2 - E)$ may then be obtained by superposition.

With this hypothesis the wave function (14.9) becomes

$$(14.17) \qquad v(t,X) = \int_K \psi_+(X,P)\, e^{-it\omega(P)}\, \hat{h}_+(P)\, dP \, .$$

The asymptotic wave function (14.5) for $v(t,X)$ will be calculated from (14.17) by substituting the expansion (14.10) and determining the behavior for $t \to +\infty$ of the terms in the resulting sum. For this purpose a bound is needed for the remainder in (14.10) that is uniform in $(p,q) \in K$. Thus a refinement of Lemma 7.3 is needed since the latter is valid for fixed p only. The following generalization of Lemma 7.3 will be proved.

<u>Lemma 14.1</u>. Define the remainder $\sigma_\pm(X,p,q)$ for all $X \in G$ and $(p,q) \in \mathcal{O}_{m,n}$ by

$$(14.18) \qquad \begin{aligned} \psi_\pm(X,p,q) &= (2\pi)^{-1} e^{i(px\ qy)} + (2\pi)^{-1} \sum_{\ell=-n}^{m} c_\ell^\pm(p,q)\, e^{i(p_\ell x \pm q_\ell y)} \\ &\quad + \sigma_\pm(X,p,q) \, . \end{aligned}$$

Then for each compact set $K \subset \mathcal{O}_{m,n}$ and each $r' > r > h$ there exist constants $\mu = \mu(K) > 0$ and $C = C(K,r',r)$ such that

$$(14.19) \qquad |\sigma_\pm(X,p,q)| \le C\, e^{-\mu y} \text{ for all } X \in R^2_{r'}, \text{ and } (p,q) \in K \, .$$

<u>Proof</u>. Only the case of σ_+ will be discussed since the other case then follows from the relation (2.25). The proof will parallel that of Lemma 7.3. Note that (8.5) implies that

$$(14.20) \qquad \sigma_+(x,y,p,q) = e^{2\pi i\ell p}\, \rho_+(x - 2\pi\ell,y,p,q) \, , \quad (x,y) \in \Omega^{(\ell)} \, ,$$

where ρ_+ is defined by (7.39) with

$$(14.21) \qquad L' = \{\ell \in Z : \ell \le -n - 1 \text{ or } \ell \ge m + 1\}$$

for all $(p,q) \in \mathcal{O}_{m,n}$. Thus to prove (14.19) it is enough to show that

$$(14.22) \qquad |\rho_+(X,p,q)| \le C\, e^{-\mu y} \text{ for all } X \in \Omega_{r'}, \text{ and } (p,q) \in K \, .$$

Proceeding as in the proof of Lemma 7.3, one has

$$(14.23) \qquad |\phi'_{+\ell}(r,p,q)|^2 \le C_0^2 (2\pi)^{-1} \|\phi'_+(\cdot,p,q)\|^2_{1;h,r} \text{ for all } \ell \in L'$$

where $C_0 = C_0(h,r)$. Now the right hand side of (14.23) is a continuous function of $(p,q) \in R_0^2 - E$ by Theorem 8.1. Then there exists a $C_1 = C_1(K,r)$ such that

$$|\phi'_{+\ell}(r,p,q)| \le C_1 \text{ for all } (p,q) \in K \text{ and } \ell \in L' . \tag{14.24}$$

Next, since K is a compact subset of $\mathcal{O}_{m,n}$ there exist constants $\mu_+ = \mu_+(K) > 0$ and $\mu_- = \mu_-(K) > 0$ such that

$$\begin{aligned} &(p + \ell)^2 - p^2 - q^2 \ge \mu_+^2 \text{ for all } (p,q) \in K \text{ and } \ell \ge m + 1 , \\ &(p + \ell)^2 - p^2 - q^2 \ge \mu_-^2 \text{ for all } (p,q) \in K \text{ and } \ell \le -n - 1 . \end{aligned} \tag{14.25}$$

whence

$$\{(p + \ell)^2 - p^2 - q^2\}^{1\ 2} \ge \mu(K) = \text{Min } (\mu_+(K),\mu_-(K)) > 0 \tag{14.26}$$

for all $(p,q) \in K$ and $\ell \in L'$. It follows that for all $X \in \Omega_{r'}$ and $(p,q) \in K$ one has

$$\begin{aligned} |\rho_+(X,p,q)| &\le \sum_{\ell\in L'} |\phi'_{+\ell}(y,p,q)| \\ &\le \sum_{\ell\in L'} |\phi'_+ (r,p,q)| \exp \{-(y-r)((p+\ell)^2-p^2-q^2)^{1/2}\} \\ &\le C_1 \sum_{\ell\in L'} \exp \{-(y-r)((p+\ell)^2-p^2-q^2)^{1/2}\} \\ &\le C_1 \sum_{\ell\in L'} \exp \{-(y-r')((p+\ell)^2-p^2-q^2)^{1/2}\} \times \\ &\quad \times \exp \{-(r'-r)((p+\ell)^2-p^2-q^2)^{1/2}\} \\ &\le C_1\, e^{-(y-r')\mu(K)} \sum_{\ell\in L'} \exp \{-(r'-r)((p+\ell)^2-p^2-q^2)^{1/2}\} . \end{aligned} \tag{14.27}$$

Now

$$\Sigma(r'-r,p,q) = \sum_{\ell\in L'} \exp \{-(r'-r)((p+\ell)^2-p^2-q^2)^{1/2}\} \tag{14.28}$$

is a continuous function of $(p,q) \in \mathcal{O}_{m,n}$ and hence for each compact $K \subset \mathcal{O}_{m,n}$ there is a constant $M(r'-r,K)$ such that

$$(14.29) \qquad \Sigma(r'-r,p,q) \le M(r'-r,K) \text{ for all } (p,q) \in K .$$

Combining (14.27), (14.28) and (14.29) gives (14.22) with

$$(14.30) \qquad C = C_1(K,r)\, e^{r'\mu(K)}\, M(r'-r,K) .$$

<u>Second Calculation of H (continued)</u>. Substitution of (14.18) into (14.17) gives the decomposition

$$(14.31) \quad v(t,X) = v^{in}(t,X) + \sum_{\ell=-n}^{m} v_\ell^{out}(t,X) + v_\sigma(t,X) \;, \; t \in R \;, \; X \in G \;,$$

where

$$(14.32) \qquad v^{in}(t,X) = \frac{1}{2\pi} \int_K e^{i(px-qy-t\omega(p,q))}\, \hat{h}_+(p,q)\, dpdq \;,$$

$$(14.33) \quad v_\ell^{out}(t,X) = \int_K e^{i(p_\ell x+q_\ell y-t\omega(p,q))}\, (2\pi)^{-1}\, c_\ell^+(p,q)\, \hat{h}_+(p,q)\, dpdq \;, \text{ and}$$

$$(14.34) \qquad v_\sigma(t,X) = \int_K e^{-it\omega(p,q)}\, \sigma_+(X,p,q)\, \hat{h}_+(p,q)\, dpdq \;.$$

Recall that by assumption $\hat{h}_+ \in C_0^\infty(R_0^2 - E)$ satisfies (14.16) and $c_\ell^+(p,q) \in C(R_0^2 - E)$ (see (11.23), (11.24)). Thus the integrands in the above integrals are all continuous. The second calculation of H will now be carried out by calculating the asymptotic wave function in $L_2(G)$ of each term on the right hand side of (14.31).

<u>The Partial Wave $v^{in}(t,X)$</u>. The change of variables $(p',q') = (p,-q)$ in (14.32) gives

$$(14.35) \quad v^{in}(t,X) = \frac{1}{2\pi} \int_{K'} e^{i(p'x+q'y-t\omega(p',q'))}\, \hat{h}_+(p',-q')\, dp'dq'$$

where $K' = \{(p',q') : (p,q) \in K\} \subset R^2 - R_0^2$. Thus $v^{in}(t,X)$ is a free wave in R^2 and hence has an asymptotic wave function $r^{-1/2} H^{in}(r-t,\theta)$ with waveform defined by [30, Theorem 2.6]

$$(14.36) \quad H^{in}(\tau,\theta) = \frac{1}{(2\pi)^{1/2}} \int_0^\infty e^{i\tau\omega}\, \hat{h}_+(\omega\cos\theta, -\omega\sin\theta)\, (-i\omega)^{1/2}\, d\omega \;.$$

In particular, $H^{in}(\tau,\theta) \equiv 0$ for $0 \leq \theta \leq \pi$ because $K' = \text{supp}\ \hat{h}_+(p,-q) \subset R^2 - R_0^2$.

<u>The Partial Waves $v_\ell^{out}(t,X)$</u>. To interpret these terms let $\ell \in Z$ and consider the mapping X_ℓ defined by

$$(14.37) \qquad (p_\ell, q_\ell) = X_\ell(p,q) = (p + \ell, \{p^2 + q^2 - (p + \ell)^2\}^{1/2}) \ .$$

X_ℓ is analytic on the domain

$$(14.38) \qquad D(X_\ell) = \{(p,q) : \sqrt{p^2 + q^2} > |p + \ell|,\ q > 0\}$$

and maps it bijectively onto the range

$$(14.39) \qquad R(X_\ell) = \{(p_\ell, q_\ell) : \sqrt{p_\ell^2 + q_\ell^2} > |p_\ell - \ell|,\ q_\ell > 0\} \ .$$

Moreover,

$$(14.40) \qquad X_\ell^{-1} = X_{-\ell}\ , \quad \ell \in Z\ ,$$

and the Jacobian of X_ℓ is

$$(14.41) \qquad \frac{\partial(p_\ell, q_\ell)}{\partial(p,q)} = \frac{q}{q_\ell}\ .$$

Note that $\omega(p,q)$ is invariant under X_ℓ :

$$(14.42) \qquad \omega(p_\ell, q_\ell) = p_\ell^2 + q_\ell^2 = p^2 + q^2 = \omega(p,q)\ .$$

It can be shown that

$$(14.43) \qquad X_\ell\, \mathcal{O}_{m,n} = \mathcal{O}_{m-\ell,n+\ell} \text{ for } -n \leq \ell \leq m\ .$$

Hence the hypothesis $K \subset \mathcal{O}_{m,n}$ implies that

$$(14.44) \qquad X_\ell K \equiv K_\ell \subset \mathcal{O}_{m-\ell,n+\ell} \text{ for } -n \leq \ell \leq m\ .$$

In particular, one has

$$(14.45) \qquad K_j \cap K_\ell = \phi \text{ for } j \neq \ell\ .$$

On making the change of variables $(p,q) \to (p_\ell, q_\ell) = X_\ell(p,q)$ in (14.33) one finds the representation

$$v_\ell^{out}(t,X) = (2\pi)^{-1} \int_{K_\ell} e^{i(xp_\ell + yq_\ell - t\omega(p_\ell,q_\ell))} \left\{ c_\ell^+(p,q)\hat{h}_+(p,q) \frac{q_\ell}{q} \right\} dp_\ell \, dq_\ell \tag{14.46}$$

where $(p,q) = X_{-\ell}(p_\ell,q_\ell)$ in the integrand. Thus $v_\ell^{out}(t,X)$ is also a free wave in R^2 and has an asymptotic wave function $r^{-1/2} H_\ell(r-t,\theta)$ with waveform defined by

$$H_\ell(\tau,\theta) = \frac{1}{(2\pi)^{1/2}} \int_0^\infty e^{i\tau\omega} \hat{H}_\ell(\omega,\theta) \, d\omega \tag{14.47}$$

and

$$\hat{H}_\ell(\omega,\theta) = (-i\omega)^{1/2} \frac{q_\ell}{q} c_\ell^+(p,q) \hat{h}_+(p,q) \Big|_{(p,q)=X_{-\ell}(\omega\cos\theta,\omega\sin\theta)} \tag{14.48}$$

A simple calculation shows that

$$\|H_\ell\|^2_{L_2(R\times[0,\pi])} = \int_K |c_\ell^+(p,q)\, \hat{h}_+(p,q)|^2 \frac{q_\ell}{q} \, dp\, dq \tag{14.49}$$

where q_ℓ is defined by (14.37). In particular, $H_\ell \in L_2(R \times [0,\pi])$ because the integrand in (14.49) is continuous on K.

The Partial Wave $v_\sigma(t,X)$. Equation (14.31) may be written

$$v_\sigma(t,X) = v(t,X) - v^{in}(t,X) - \sum_{\ell=-n}^{m} v_\ell^{out}(t,X) \tag{14.50}$$

for all $t \in R$ and $X \in G$. Moreover, it has been shown that

$$\begin{aligned} v(t,X) &= r^{-1/2} H(r-t,\theta) + o(1) , \\ v^{in}(t,X) &= o(1) , \\ v_\ell^{out}(t,X) &= r^{-1/2} H_\ell(r-t,\theta) + o(1) \end{aligned} \tag{14.51}$$

where each term $o(1) \in L_2(G)$ for all $t \in R$ and tends to zero in $L_2(G)$ when $t \to +\infty$. These results imply that

$$v_\sigma(t,X) = r^{-1/2} H_\sigma(r-t,\theta) + o(1) \tag{14.52}$$

where $o(1) \to 0$ in $L_2(G)$ when $t \to +\infty$ and

$$(14.53) \qquad H_\sigma(\tau,\theta) = H(\tau,\theta) - \sum_{\ell=-n}^{m} H_\ell(\tau,\theta) \text{ in } L_2(R \times [0,\pi]) .$$

On the other hand (15.34) and Lemma 14.1 imply that

$$(14.54) \qquad |v_\sigma(t,X)| \leq C_\sigma e^{-\mu y} \text{ for all } t \in R \text{ and } X \in R^2_{r'},$$

where

$$(14.55) \qquad C_\sigma = C \int_K |\hat{h}_+(p,q)| \, dpdq$$

and $\mu = \mu(K) > 0$ and $C = C(K,r',r)$ are the constants of the lemma. The second calculation of H will now be completed by showing that (14.52) and (14.54) imply

Theorem 14.2. $H_\sigma(\tau,\theta) \equiv 0$ and hence

$$(14.56) \qquad H(\tau,\theta) = \sum_{\ell=-n}^{m} H_\ell(\tau,\theta) \text{ in } L_2(R \times [0,\pi]) .$$

Proof. Let ε be an arbitrary number in the interval $0 < \varepsilon < \pi/2$ and consider the sector

$$(14.57) \qquad \Gamma_\varepsilon = \{(x,y) = (r \cos \theta, r \sin \theta) : \varepsilon < \theta < \pi - \varepsilon\} .$$

By (14.52), the local decay of asymptotic wave functions [30, p. 32] and the triangle inequality, one has

$$(14.58) \qquad \begin{aligned} \int_{G\cap\Gamma_\varepsilon} |v_\sigma(t,X)|^2 \, dX &= \int_{\Gamma_\varepsilon} |H_\sigma(r-t,\theta)|^2 r^{-1} \, dX + o(1) \\ &= \int_0^\infty \int_\varepsilon^{\pi-\varepsilon} |H_\sigma(r-t,\theta)|^2 \, d\theta \, dr + o(1) \\ &= \int_{-t}^\infty \int_\varepsilon^{\pi-\varepsilon} |H_\sigma(\tau,\theta)|^2 \, d\theta \, d\tau + o(1) \end{aligned}$$

where $o(1) \to 0$ when $t \to \infty$. Thus passage to the limit in (15.58) gives

$$(14.59) \qquad \lim_{t\to+\infty} \int_{G\cap\Gamma_\varepsilon} |v_\sigma(t,X)|^2 \, dX = \int_{-\infty}^\infty \int_\varepsilon^{\pi-\varepsilon} |H_\sigma(\tau,\theta)|^2 \, d\theta \, d\tau .$$

On the other hand, writing $R^2_{a,b} = \{(x,y) : x \in R, a < y < b\}$, one has

$$\int_{G\cap\Gamma_\varepsilon} |v_\sigma(t,X)|^2\, dX = \int_{G\cap\Gamma_\varepsilon\cap R^2_{0,k}} |v_\sigma(t,X)|^2\, dX + \int_{\Gamma_\varepsilon\cap R^2_k} |v_\sigma(t,X)|^2\, dX$$

(14.60)

$$\leq C_\sigma^2 \int_{\Gamma_\varepsilon\cap R^2_k} e^{-2\mu r \sin\theta}\, r\, dr\, d\theta + o(1)$$

for every fixed $k > r' > r$ by (14.54) and the local decay property for v_σ. Passage to the limit $t \to +\infty$ in (14.60) gives by (14.59),

$$(14.61)\qquad \int_{-\infty}^{\infty}\int_{\varepsilon}^{\pi-\varepsilon} |H_\sigma(\tau,\theta)|\, d\theta\, d\tau \leq C_\sigma^2 \int_{\Gamma_\varepsilon\cap R^2_k} e^{-2\mu r\sin\theta}\, r\, dr\, d\theta$$

for every $k > r'$. Note that C_σ, μ and the left hand side of (14.61) are independent of k. Now, $\sin\theta \geq \sin\varepsilon > 0$ for $\varepsilon \leq \theta \leq \pi - \varepsilon$ and hence

$$\int_{\Gamma_\varepsilon\cap R^2_k} e^{-2\mu r\sin\theta}\, r\, dr\, d\theta \leq \int_{\Gamma_\varepsilon\cap R^2_k} e^{-2\mu r\sin\varepsilon}\, r\, dr\, d\theta$$

(14.62)

$$\leq \int_k^\infty \int_\varepsilon^{\pi-\varepsilon} e^{-2\mu r\sin\varepsilon}\, r\, d\theta\, dr$$

$$= (\pi - 2\varepsilon)\int_k^\infty e^{-2\mu r\sin\varepsilon}\, r\, dr\ .$$

But the last integral tends to zero when $k \to \infty$ with ε fixed. Thus (14.61) implies that $H_\sigma(\tau,\theta) \equiv 0$ in $R \times [\varepsilon,\pi-\varepsilon]$ and (14.56) follows since $\varepsilon \in [0,\pi/2]$ is arbitrary.

Corollary 14.3. For all $h \in L_2(G)$ such that $\operatorname{supp}\hat{h}_\pm \subset \bar{\mathcal{O}}_{m,n}$ = closure of $\mathcal{O}_{m,n}$ one has the two relations

$$(14.63)\qquad \hat{h}_\pm(p,q) = \sum_{\ell=-n}^{m} c_\ell^\pm(X_{-\ell}(p,q))\, \hat{h}_\pm(X_{-\ell}(p,q))\, \frac{q}{q_{-\ell}}$$

for almost every $(p,q) \in R_0^2$ where $(p_{-\ell},q_{-\ell}) = X_{-\ell}(p,q)$.

Proof. The case where $\operatorname{supp}\hat{h}_+ \subset \bar{\mathcal{O}}_{m,n}$ is considered first. In this case it will suffice to prove (14.63) for functions $h \in L_2(G)$ such that $\hat{h}_+ \in C_0^\infty(R_0^2 - \varepsilon)$ and $\operatorname{supp}\hat{h}_+ = K \subset \mathcal{O}_{m,n}$ since such functions are dense in the subspace of $L_2(G)$ defined by $\operatorname{supp}\hat{h}_+ \subset \bar{\mathcal{O}}_{m,n}$. For such functions the relation (14.56) and the Fourier representations (14.8) for $H(\tau,\theta)$ and (14.47), (14.48) for $H_\ell(\tau,\theta)$ imply that

$$\hat{h}_-(\omega\cos\theta,\omega\sin\theta) = \sum_{\ell=-n}^{m} c_\ell^+(X_{-\ell}(\omega\cos\theta,\omega\sin\theta)) \times \tag{14.64}$$

$$\times\ \hat{h}_+(X_{-\ell}(\omega\cos\theta,\omega\sin\theta)) \times$$

$$\times\ \frac{\omega\sin\theta}{\{\omega^2-(\omega\cos\theta-\ell)^2\}^{1/2}}$$

for almost every $(\omega,\theta) \in R_0 \times [0,\pi]$. Making the substitutions $p = \omega\cos\theta$, $q = \omega\sin\theta$ in (14.64) gives (14.63) in the case $\mathrm{supp}\ \hat{h}_+ \subset \mathcal{O}_{m,n}$.

The second case of (14.63) can be derived by calculating the asymptotic wave functions for $v(t,\cdot) = e^{-itA^{1/2}} h$ when $t \to -\infty$, using the method given above. A simpler derivation may be based on the relations $\psi_-(X,p,q) = \overline{\psi_+(X,-p,q)}$ and $c_\ell^-(p,q) = \overline{c_{-\ell}^+(-p,q)}$ of (2.25) and (2.29). Indeed, if $\mathrm{supp}\ \hat{h}_- \subset \mathcal{O}_{m,n}$ and $g(X) = \overline{h(X)}$ then these relations imply that $\hat{g}_\pm(p,q) = \overline{\hat{h}_\mp(p,q)}$ and hence relation (14.63) with the upper sign for g implies (14.63) with the lower sign for h.

The Structure of $\hat{S}$. It will be convenient to use the notation

$$g_{m,n}(P) = \chi_{m,n}(P)\ g(P) \tag{14.65}$$

where $\chi_{m,n}$ is the characteristic function of the set $\mathcal{O}_{m,n}$. Clearly, the operator $P_{m,n}$ in $L_2(R_0^2)$ defined by

$$P_{m,n}\ g = g_{m,n}\ , \quad m,n = 0,1,2,\cdots\ , \tag{14.66}$$

is an orthogonal projection and different operators of the family have orthogonal ranges. Moreover, the relation (14.15) implies that the family is complete because E is a null set; i.e.,

$$\sum_{m,n=0}^{\infty} P_{m,n} = 1\ . \tag{14.67}$$

It follows that for all $g \in L_2(R_0^2)$ one has

$$\hat{S}\ g = \sum_{m,n=0}^{\infty} \hat{S}(g_{m,n})\ . \tag{14.68}$$

Thus $\hat{S}$ is completely determined by

Theorem 14.4. For all $g \in L_2(R_0^2)$ one has

$$\hat{S}(g_{m,n}) = \sum_{\ell=-n}^{m} (\hat{S}\, g_{m,n})_{m-\ell,n+\ell}\ , \text{ and} \tag{14.69}$$

$$(\hat{S}_{m,n})_{m-\ell,n+\ell}(p,q) = \frac{q}{q_{-\ell}}\, c_\ell^+(X_{-\ell}(p,q))\ g_{m,n}(X_{-\ell}(p,q))\ . \tag{14.70}$$

Similarly, one has

$$\hat{S}^*(g_{m,n}) = \sum_{\ell=-n}^{m} (\hat{S}^*\, g_{m,n})_{m-\ell,n+\ell}\ , \text{ and} \tag{14.71}$$

$$(\hat{S}^*\, g_{m,n})_{m-\ell,n+\ell}(p,q) = \frac{q}{q_{-\ell}}\, c_\ell^-(X_{-\ell}(p,q))\ g_{m,n}(X_{-\ell}(p,q))\ . \tag{14.72}$$

In particular, if supp $g \subset \overline{O}_{m,n}$ then

$$\text{supp}\ \hat{S}\ g \cup \text{supp}\ \hat{S}^* \subset \bigcup_{\ell=-n}^{m} \overline{O}_{m-\ell,n+\ell}\ . \tag{14.73}$$

Proof. Equations (14.69), (14.70), (14.71) and (14.72) follow immediately from Corollary 14.3, the relations $\hat{h}_- = \hat{S}\ \hat{h}_+$, $\hat{h}_+ = \hat{S}^*\ \hat{h}_-$ and the observation that when supp $\hat{h}_+ \subset \overline{O}_{m,n}$ then the ℓth term in the sum in (14.63) has its support in $\overline{O}_{m-\ell,n+\ell}$. (14.73) follows from (14.69) and (14.71).

The unitarity of $\hat{S}$ and (14.70), (14.72) impose restrictions on the scattering coefficients $c_\ell^\pm$. To calculate them it will be convenient to calculate $\hat{S}^*$ and $\hat{S}$ directly from (14.70) and (14.72), respectively. This gives the following alternative representations of $\hat{S}^*$ and $\hat{S}$.

Theorem 14.5. For all $g \in L_2(R_0^2)$ one has

$$(\hat{S}^*\, g_{m,n})_{m-\ell,n+\ell}(p,q) = \overline{c_{-\ell}^+(p,q)}\ g_{m,n}(X_{-\ell}(p,q)) \tag{14.74}$$

and similarly

$$(\hat{S}\, g_{m,n})_{m-\ell,n+\ell}(p,q) = \overline{c_{-\ell}^-(p,q)}\ g_{m,n}(X_{-\ell}(p,q))\ . \tag{14.75}$$

Proof. For all $f,g \in L_2(R_0^2)$ one has

$$(f,(\hat{S}^*\, g)_{m,n}) = (f_{m,n},\hat{S}^*\, g) = (\hat{S}(f_{m,n}),\ g) = \sum_{\ell=-n}^{m} (\hat{S}(f_{m,n}),g_{m-\ell,n+\ell}) \tag{14.76}$$

$$(14.76 \text{ cont.}) \qquad = \sum_{\ell=-n}^{m} \int_{\mathcal{O}_{m-\ell,n+\ell}} \overline{c_\ell^{\mp}(X_{-\ell})\; f_{m,n}(X_{-\ell})}\; g_{m-\ell,n+\ell}\, \frac{q}{q_{-\ell}}\, dp\,dq$$

by (14.70). On making the change of variables $(p',q') = (p_{-\ell},q_{-\ell}) = X_{-\ell}(p,q)$ in the last integral and noting that $q/q_{-\ell} = \partial(p',q')/\partial(p,q)$, one has

$$(14.77) \qquad \begin{aligned} (f,(\hat{S}^* g)_{m,n}) &= \sum_{\ell=-n}^{m} \int_{\mathcal{O}_{m,n}} \overline{c_\ell^{\mp}(p',q') f_{m,n}(p',q')}\; g_{m-\ell,n+\ell}(X_\ell(p',q'))\, dp'dq' \\ &= \int_{R_0^2} \overline{f(p,q)} \left[\sum_{\ell=-n}^{m} \overline{c_\ell^{\mp}(p,q)}\; g_{m-\ell,n+\ell}(X_\ell(p,q)) \right] dp\,dq \end{aligned}$$

because $\operatorname{supp} g_{m-\ell,n+\ell}(X_\ell) \subset \overline{\mathcal{O}}_{m,n}$. Since $f \in L_2(R_0^2)$ is arbitrary, (14.77) implies that

$$(14.78) \qquad (\hat{S}^* g)_{m,n}(p,q) = \sum_{\ell=-n}^{m} \overline{c_\ell^{\mp}(p,q)}\; g_{m-\ell,n+\ell}(X_\ell(p,q)) \; .$$

To derive (14.74) note that for all ℓ, m, n, $\bar{m}$, $\bar{n} \geq 0$

$$(14.79) \qquad (g_{\bar{m},\bar{n}})_{m-\ell,n+\ell} = \delta_{m-\ell,\bar{m}}\, \delta_{n+\ell,\bar{n}}\, g_{\bar{m},\bar{n}}$$

where δ_{jk} is the Kronecker symbol. Noting that $\delta_{m-\ell,\bar{m}}\, \delta_{n+\ell,\bar{n}} = \delta_{\ell,m-\bar{m}}\, \delta_{\ell,\bar{n}-n} = \delta_{m-\bar{m},\bar{n}-n}\, \delta_{\ell,m-\bar{m}}$, (14.78) and (14.79) imply

$$(14.80) \qquad \begin{aligned} (\hat{S}^* g_{\bar{m},\bar{n}})_{m,n}(p,q) &= \delta_{m-\bar{m},\bar{n}-n} \sum_{\ell=-n}^{m} \overline{c_\ell^{\mp}(p,q)}\; \delta_{\ell,m-\bar{m}}\, g_{\bar{m},\bar{n}}(X_\ell(p,q)) \\ &= \delta_{m-\bar{m},\bar{n}-n}\, \overline{c_{m-\bar{m}}^{\mp}(p,q)}\; g_{\bar{m},\bar{n}}(X_{m-\bar{m}}(p,q)) \; . \end{aligned}$$

This clearly is zero unless $\bar{m} - m = n - \bar{n} = \ell$ where $-n \leq \ell \leq m$, which implies (14.71). Moreover, setting $(m,n) = (\bar{m} - \ell, \bar{n} + \ell)$ in (14.80) gives (14.74). The proof of (14.75) is obtained by the same method, beginning with (14.72).

The two representations of $\hat{S}$ and $\hat{S}^*$ of Theorems 14.4 and 14.5 hold for arbitrary $g \in L_2(R_0^2)$. It follows that the scattering coefficients must satisfy the relations

$$(14.81) \qquad q\, c_\ell^{\pm}(X_{-\ell}(p,q)) = q_{-\ell}\, \overline{c_{-\ell}^{\mp}(p,q)}$$

for all $(p,q) \in \mathcal{O}_{m-\ell,n+\ell}$. Moreover, the unitarity of $\hat{S}$ and Theorems 14.4 and 14.5 imply

Theorem 14.6. The scattering coefficients $c_\ell^\pm$ of the R-B waves $\psi_\pm(X,p,q)$ satisfy the identities

$$\sum_{\ell=-n}^{m} \overline{c_\ell^\pm(p,q)}\; c_{\ell-k}^\pm\,(X_k(p,q))q_\ell = q\,\delta_{k,0}\ , \text{ and} \tag{14.82}$$

$$\sum_{\ell=-n}^{m} \overline{c_{-\ell}^\pm(X_\ell(p,q))}\; c_{k-\ell}^\pm(X_\ell(p,q))q_\ell^{-1} = q^{-1}\,\delta_{k,0} \tag{14.83}$$

for all $(p,q) \in \mathcal{O}_{m,n}$ and all k such that $-n \le k \le m$.

These properties may be verified by simple calculations using the relations

$$(\hat{S}(f_{m,n}),\hat{S}(g_{m-k,n+k})) = \delta_{k,0}(f_{m,n},g_{m,n})\ , \text{ and} \tag{14.84}$$

$$(\hat{S}^*(f_{m,n}),\hat{S}^*(g_{m-k,n+k})) = \delta_{k,0}(f_{m,n},g_{m,n}) \tag{14.85}$$

and the constructions of $\hat{S}$ and $\hat{S}^*$ described in Theorems 14.4 and 14.5. Relation (14.83) also follows from relations (14.81) and (14.82).

It is well known in the theories of scattering by potentials and by bounded obstacles that the S-matrix $\hat{S}$ is a direct integral of a family of unitary operators $\hat{S}(\omega)$ that act on the "energy shell" $p^2 + q^2 = \omega^2$. The analogous property of the S-matrices for diffraction gratings is evident from Theorem 14.4 and the properties of the mappings X_ℓ. The operator $\hat{S}(\omega)$ in this case is given by (cf. (14.75))

$$\hat{S}(\omega)\ g(\omega\cos\theta,\omega\sin\theta) = \sum_{\ell=-m}^{n} \overline{c_\ell^-(\omega\cos\theta,\omega\sin\theta)}\ g(X_\ell(\omega\cos\theta,\omega\sin\theta)) \tag{14.86}$$

when $\operatorname{supp} g \subset \overline{\mathcal{O}}_{m,n}$. If $s(\theta) = g(\omega\cos\theta,\omega\sin\theta)$ is an arbitrary function with $\operatorname{supp} s \subset \{\theta : (\omega\cos\theta,\omega\sin\theta) \in \overline{\mathcal{O}}_{m,n}\}$ then (14.86) can be written

$$(\hat{S}(\omega)s)(\theta) = \sum_{\ell=-m}^{n} \overline{c_\ell^-(\omega\cos\theta,\omega\sin\theta)}\ s(\theta_\ell) \tag{14.87}$$

where $\theta_\ell = \theta_\ell(\omega,\theta)$ is defined as the unique angle such that $0 \le \theta_\ell \le \pi$ and

$$X_\ell(\omega\cos\theta,\omega\sin\theta) = (\omega\cos\theta_\ell,\omega\sin\theta_\ell)\ . \tag{14.88}$$

For general $s \in L_2(0,\pi)$, $\hat{S}(\omega)s$ is obtained from (14.87), (14.88) by superposition. The unitarity of $\hat{S}(\omega)$ in $L_2(0,\pi)$ can be verified by direct calculation using (14.87), the analogue for $\hat{S}^*(\omega)$ and Theorem 14.6.

§15. The Scattering of Signals by Diffraction Gratings

The results of §13 and §14 are applicable to the echoes that are produced when signals generated by localized sources are scattered by a diffraction grating. The structure of such echoes is analyzed in this section. Most of the section deals with the case, often realized in applications, of sources that are far from the grating. In particular, it is shown that is this case the influence of the grating on the echoes is completely described by the S-matrix.

It will be assumed that the sources of the signals are localized near a point $(0,y_0) \in G$ and act during a time interval $T \le t \le 0$. The resulting wave field $u(t,X)$ is then characterized by its initial values $u(0,X)$, $D_t u(0,X)$ in G. To make explicit their dependence on y_0 the initial values will be assumed to have the form

$$(15.1)\qquad \begin{aligned} u(0,X) &= f(X,y_0) \equiv f_0(x,y-y_0) , \\ D_t u(0,X) &= g(X,y_0) \equiv g_0(x,y-y_0) \end{aligned}$$

for all $X = (x,y) \in G$ where $y_0 \ge 0$,

$$(15.2)\qquad f_0 \in L_2^{1,com}(G) , \quad g_0 \in L_2^{com}(G) ,$$

and $f(X,y_0) \equiv g(X,y_0) \equiv 0$ for $(x,y-y_0) \notin G$. Note that for $y_0 \ge 0$ one has $f(\cdot,y_0) \in D(A^{1/2})$, $g(\cdot,y_0) \in L_2(G)$ and hence $u(t,X)$ is a solution wFE in G. The functions $f(\cdot,y_0)$, $g(\cdot,y_0)$ will also be used as initial values for free waves in R^2 and for wave fields in the degenerate grating domain R_0^2. In each case the domain under consideration will be clear from the context or will be stated explicitly. For brevity, the coordinate y_0 will be suppressed except in places where the y_0-dependence is under discussion.

The Signal Wave Field. In the absence of a diffraction grating the initial state f, g will generate a signal wave field $u_s(t,X)$ in R^2. The first derivatives of $u_s(t,X)$ have asymptotic wave functions [30, Theorem 2.10]

$$(15.3)\qquad D_k\, u_s(t,X) = r^{-1/2}\, s_k(r-t,\theta) + o(1) , \quad k = 0,1,2 ,$$

where the waveforms $s_k(\tau,\theta) \in L_2(R \times [-\pi,\pi])$ are given by

$$(15.4) \quad s_0(\tau,\theta) = \text{Re}\left\{\frac{1}{(2\pi)^{1/2}}\int_0^\infty e^{i\tau\omega}\,\hat{h}(\omega\cos\theta,\omega\sin\theta)(-i\omega)^{1/2}\,d\omega\right\},$$

$s_1(\tau,\theta) = -s_0(\tau,\theta)\cos\theta$, $s_2(\tau,\theta) = -s_0(\tau,\theta)\sin\theta$ and the terms $o(1) \to 0$ in $L_2(R^2)$ when $t \to \infty$. The function

$$(15.5) \qquad \hat{h}(P) = \Phi\, g(P) - i\omega(P)\,\Phi\, f(P)$$

where Φ denotes the Fourier transform in $L_2(R^2)$. In particular, the Fourier transform in $L_2(R \times [-\pi,\pi])$ of the signal waveform $s_0(\tau,\theta)$ is

$$(15.6) \qquad \hat{s}_0(\omega,\theta) = \frac{1}{2}(-i\omega)^{1/2}\,\hat{h}(\omega\cos\theta,\omega\sin\theta)\,.$$

It can be verified that if f and g are real-valued then $\hat{s}_0(-\omega,\theta) = \overline{\hat{s}_0(\omega,\theta)}$ and hence (15.6) generates a real-valued signal.

When y_0 is large the signal arriving at the grating surface is described by the signal waveform $s_0(\tau,\theta)$ through (15.3). The problem of signal design is to construct a source or "transmitter" whose waveform $s_0(\tau,\theta)$ approximates a prescribed function. The solution of this problem is the task of the transmitter design engineer.

<u>The Echo Wave Fields</u>. In the presence of a diffraction grating with domain G the initial state f,g will generate a total wave field u(t,X) whose asymptotic behavior for $t \to +\infty$ is described by Theorem 13.3. In particular,

$$(15.7) \qquad D_0 u(t,X) = r^{-1/2} F_0(r-t,\theta) + o(1) \text{ in } L_2(G)\,, \quad t \to +\infty\,,$$

where $F_0(\tau,\theta) \in L_2(R \times [0,\pi])$ is defined by

$$(15.8) \quad F_0(\tau,\theta) = \text{Re}\left\{\frac{1}{(2\pi)^{1/2}}\int_0^\infty e^{i\tau\omega}\,\hat{h}_-(\omega\cos\theta,\omega\sin\theta)(-i\omega)^{1/2}\,d\omega\right\}$$

and

$$(15.9) \qquad \hat{h}_-(p) = \hat{g}_-(P) - i\omega(P)\,\hat{f}_-(P)\,.$$

The echo wave field $u_e(t,X)$ is defined by

$$(15.10) \qquad u_e(t,X) = u(t,X) - u_s(t,X)\,, \quad t \geq 0\,, \quad X \in G\,.$$

Thus the echo is described for large t by

$$(15.11) \qquad D_0 u_e(t,X) = r^{-1/2} e_0(r-t,\theta) + o(1) \text{ in } L_2(G) \ , \quad t \to +\infty \ ,$$

where $e_0 = F_0 - s_0 \in L_2(R \times [0,\pi])$ is given by

$$(15.12) \qquad e_0(\tau,\theta) = \operatorname{Re}\left\{\frac{1}{(2\pi)^{1/2}} \int_0^\infty e^{i\tau\omega}\, \hat{h}_-^{sc}(\omega \cos\theta, \omega \sin\theta)(-i\omega)^{1/2}\, d\omega\right\}$$

with

$$(15.13) \qquad \hat{h}_-^{sc}(P) = \hat{h}_-(P) - \hat{h}(P) = \hat{g}_-^{sc}(P) - i\omega(P)\, \hat{f}_-^{sc}(P) \ .$$

The last functions can be written in terms of the R-B diffracted plane waves (see (2.24))

$$(15.14) \qquad \psi_-(X,P) = \psi^{inc}(X,p,-q) + \psi_-^{sc}(X,P)$$

as

$$(15.15) \qquad \hat{g}_-^{sc}(P) = \int_G \overline{\psi_-^{sc}(X,P)}\, g(X)\, dX$$

with the analogous representation for $\hat{f}_-^{sc}$.

<u>The Echoes of Signals from Remote Sources</u>. Equations (15.12)-(15.15) provide a construction of the echo due to an arbitrary distribution of sources. The principal goal of this section is to determine how this construction may be simplified when the sources are far from the grating; i.e., $y_0 \to \infty$. To this end recall the decomposition of Lemma 14.1. Substituting equation (14.18) in (15.15) gives

$$(15.16) \qquad \hat{g}_-^{sc}(p,q) = \sum_{\ell=-n}^{m} \overline{c_\ell^-(p,q)}\, \hat{g}(p_\ell,-q_\ell) + \rho_{m,n}(p,q) \ , \quad (p,q) \in \mathcal{O}_{m,n} \ ,$$

where $\hat{g} = \Phi\, g$ is the Fourier transform in $L_2(R^2)$ and

$$(15.17) \qquad \rho(p,q) \equiv \rho(p,q;g) = \int_G \overline{\sigma_-(X,p,q)}\, g(X)\, dX \ , \quad (p,q) \in R_0^2 - E \ .$$

Note that if the unitary operator $R : L_2(R^2) \to L_2(R^2)$ is defined by

$$(15.18) \qquad R\, f(x,y) = f(x,-y)$$

then

$$(15.19) \qquad (\Phi\, R\, f)(p,q) = (R\, \hat{f})(p,q) = \hat{f}(p,-q) \; .$$

Hence (15.16) implies that for all $(p,q) \in \mathcal{O}_{m,n}$ one has

$$(15.20) \qquad \begin{aligned} \hat{g}_-^{sc}(P) &= \sum_{\ell=-n}^{m} \overline{c_\ell^-(P)}\, (R\, \hat{g})(P_\ell) + \rho_{m,n}(P) \\ &= \sum_{\ell=-m}^{n} \overline{c_{-\ell}^-(P)}\, (R\, \hat{g})_{m+\ell,n-\ell}(X_{-\ell}(P)) + \rho_{m,n}(P) \\ &= \sum_{\ell=-m}^{n} (\hat{S}(R\, \hat{g})_{m+\ell,n-\ell})_{m,n}(P) + \rho_{m,n}(P) \\ &= (\hat{S}\, R\, \hat{g})_{m,n}(P) + \rho_{m,n}(P) \end{aligned}$$

by Theorems 14.4 and 14.5. Proceeding in the same way with $\omega(P)\, \hat{f}_-^{sc}(P)$ and recalling that $\omega(P) = \omega(P_\ell)$ one finds

$$(15.21) \qquad \hat{h}_-^{sc}(P) = (\hat{S}\, R\, \hat{h})(P) + \rho(P;h) \; , \quad P \in R_0^2 - E \; ,$$

where

$$(15.22) \qquad \rho(P;h) = \rho(P;g) - i\omega(P)\, \rho(P;f) \; .$$

The estimate (14.19) of Lemma 14.1 clearly implies that $\rho(P;h(\cdot,y_0)) \to 0$ when $y_0 \to \infty$, uniformly for P in any compact subset of $R_0^2 - E$. This result is not strong enough to yield a corresponding estimate of the echo waveform $e_0(\tau,\theta)$ defined by (15.12) and it is natural to conjecture that $\rho(\cdot;h(\cdot,y_0)) \to 0$ in $L_2(R_0^2)$ when $y_0 \to \infty$. Unfortunately, if one assumes only that $A(G)$ admits no surface waves then this property does not not follow from the results obtained above because no information was obtained concerning the behavior of $\psi_-(X,P)$ for P near the exceptional set E. However, in those cases where the analytic continuation of the resolvent of A_p has no singularities on $\sigma(A_p)$ (i.e., $\Sigma_p \cap \sigma(A_p) = \phi$ for every $p \in (-1/2,1/2]$) the limiting absorption theorem, Corollary 4.17, is valid on all of $\sigma(A_p)$ (see Theorem 4.15) and Theorem 8.1 can be improved to state that $\psi_\pm(\cdot,P)$ exists and $P \to \psi_\pm(\cdot,P) \in L_2^{1,\ell oc}(\Delta,G)$ is continuous for all $P \in \overline{R_0^2}$. This improvement of Theorem 8.1 implies

Theorem 15.1. Let A(G) have no surface waves and, in addition, assume that

$$\Sigma_p \cap \sigma(A_p) = \phi \text{ for all } p \in (-1/2,1/2] \;. \tag{15.23}$$

Then for every $g_0 \in L_2(G)$ one has

$$\hat{g}_-^{sc}(\cdot,y_0) = \hat{S}\ R\ \hat{g}(\cdot,y_0) + o(1) \text{ in } L_2(R_0^2)\;,\quad y_0 \to \infty\;. \tag{15.24}$$

Similarly, for all $f_0 \in L_2^1(G)$ one has

$$\omega(\cdot)\ \hat{f}_-^{sc}(\cdot,y_0) = \omega(\cdot)\ \hat{S}\ R\ \hat{f}(\cdot,y_0) + o(1) \text{ in } L_2(R_0^2)\;,\quad y_0 \to \infty\;. \tag{15.25}$$

The proof of Theorem 15.1 will be based on the following extension of Lemma 14.1.

Lemma 15.2. Under the hypotheses of Theorem 15.1, for every compact set $K \subset \overline{R_0^2}$ and every $r' > r > h$ there is a constant $C = C(K,r,r')$ such that

$$|\sigma_\pm(X,P)| \le C \text{ for all } X \in R^2_{r'} \text{ and } P \in K\;. \tag{15.26}$$

Proof of Lemma 15.2. It clearly suffices to prove the lemma for the case $K = \overline{\mathcal{O}}_{m,n}$. On examining the proof of Lemma 14.1 one finds that the continuity of $P \to \psi_\pm(\cdot,P)$ for all $P \in \overline{R_0^2}$ implies that (14.24) holds for $K = \overline{\mathcal{O}}_{m,n}$. Moreover, (14.25) holds for all $P \in \overline{\mathcal{O}}_{m,n}$ with $\mu_+ = \mu_- = 0$. Thus (15.26) follows from (14.27) with $\mu(K) = 0$.

Proof of Theorem 15.1. Note first that if the translation operator $T_{y_0} : L_2(G) \to L_2(G)$ is defined for each $y_0 \ge 0$ by $T_{y_0} g_0 = g(\cdot,y_0)$ then (15.24) is equivalent to the statement that

$$\operatorname*{s-lim}_{y_0\to\infty}\ (\Phi_- - \Phi - \hat{S}\ R\ \Phi)T_{y_0} = 0\;. \tag{15.27}$$

Moreover, the family of operators in (15.27) is uniformly bounded for all $y_0 \ge 0$. Hence by a familiar density argument (cf. [30, proof of Theorem 2.6]) it will suffice to establish (15.24) for all g_0 in a dense subset of $L_2(G)$. The set $C_0^\infty(G)$ will be chosen for this purpose. Thus the proof will be completed by showing that if $g_0 \in C_0^\infty(G)$ and

$$\rho(P,g(\cdot,y_0)) = \hat{g}_-^{sc}(P,y_0) - \hat{S}\ R\ \hat{g}(\cdot,y_0)(P) = \int_G \overline{\sigma_-(X,P)}\ g(X,y_0)dX\;,\quad P \in R_0^2\;, \tag{15.28}$$

then

$$\lim_{y_0\to\infty} \int_{R_0^2} |\rho(P,g(\cdot,y_0))|^2 \, dP = 0 \; . \tag{15.29}$$

To prove (15.29) it will be convenient to decompose R_0^2 as the disjoint union

$$R_0^2 = D(\gamma) \cup (D'(\gamma) \cap E_\delta) \cup (D'(\gamma) - E_\delta) \tag{15.30}$$

where

$$\begin{aligned} D(\gamma) &= R_0^2 \cap \{P : |P| \geq \gamma\} , \\ D'(\gamma) &= R_0^2 \cap \{P : |P| \leq \gamma\} , \text{ and} \\ E_\delta &= R_0^2 \cap \{P : \text{dist } (P,E) \leq \delta\} . \end{aligned} \tag{15.31}$$

With this notation the integral in (15.29) can be written

$$\int_{R_0^2} |\rho(P,g(\cdot,y_0))|^2 \, dP = I_1(\gamma,y_0) + I_2(\gamma,\delta,y_0) + I_3(\gamma,\delta,y_0) \tag{15.32}$$

where

$$I_1(\gamma,y_0) = \int_{D(\gamma)} |\rho(P,g(\cdot,y_0)|^2 \, dP \; , \tag{15.33}$$

$$I_2(\gamma,\delta,y_0) = \int_{D'(\gamma)\cap E_\delta} |\rho(P,g(\cdot,y_0))|^2 \, dP \; , \text{ and} \tag{15.34}$$

$$I_3(\gamma,\delta,y_0) = \int_{D'(\gamma)-E_\delta} |\rho(P,g(\cdot,y_0))|^2 \, dP \; . \tag{15.35}$$

To estimate $I_1(\gamma,y_0)$ note that (14.18) implies that $(\Delta + |P|^2)\, \sigma_-(X,P) = 0$ for all $X \in G$, $P \in R_0^2$. Thus integrating by parts in (15.28) gives

$$\begin{aligned} (P,g(\cdot,y_0)) &= -|P|^{-2} \int_G \overline{\Delta\sigma_-(X,P)} \, g(X,y_0) \, dX \\ &= -|P|^{-2} \int_G \overline{\sigma_-(X,P)} \, \Delta g(X,y_0) \, dX \\ &= -|P|^{-2} \, \{\Delta g(\cdot,y_0)^{\wedge,sc}_-(P) - \hat{S} \, R(\Delta g(\cdot,y_0))^\wedge(p)\} \\ &= -|P|^{-2} \, \{(\Delta g(\cdot,y_0))^\wedge_-(P) - (\Delta g(\cdot,y_0))^\wedge(P) - \hat{S}R(\Delta g(\cdot,y_0))^\wedge(P)\} \end{aligned} \tag{15.36}$$

On squaring (15.36), integrating over $D(\gamma)$ and using the inequality $|z_1 + z_2 + z_3|^2 \le 4(|z_1|^2 + |z_2|^2 + |z_3|^2)$ one finds

$$I_1(\gamma,y_0) \le \int_{D(\gamma)} |P|^{-4} \, |(\Delta g)\hat{}_- - (\Delta g)\hat{} - \hat{S}\, R(\Delta g)\hat{}|^2 \, dP$$

(15.37)

$$\le 4\gamma^{-4}(\|\Phi_-(\Delta g(\cdot,y_0))\|^2 + \|\Phi(\Delta g(\cdot,y_0)\|^2 + \|\hat{S}\, R(\Delta g(\cdot,y_0))\|^2)$$

$$\le 12\gamma^{-4} \, \|\Delta g(\cdot,y_0)\|^2 = 12\gamma^{-4} \, \|\Delta g_0\|^2_{L_2(G)}$$

for all $y_0 \ge 0$. In particular, $I_1(\gamma,y_0)$ is small for large γ, uniformly in $y_0 \ge 0$.

Now consider $I_2(\gamma,\delta,y_0)$. Lemma 15.2 and equation (15.28) imply that for all $P \in D'(\gamma)$ one has

$$|\rho(P,g(\cdot,y_0)| \le C(D'(\gamma),r,r') \int_G |g_0(x,y-y_0)| \, dxdy$$

(15.38)

$$= C_1(\gamma,r,r') \int_G |g_0(X)| \, dX = C_2(g_0,\gamma,r,r') \; .$$

Combining this and (15.34) gives

$$I_2(\gamma,\delta,y_0) \le C_2^2(g_0,\gamma,r,r') \, |D'(\gamma) \cap E_\delta| \; , \tag{15.39}$$

for all $y_0 \ge 0$, where $|M|$ denotes the Lebesgue measure of a set $M \subset R^2$. Finally, note that Lemma 14.1 implies that $\rho(P,g(\cdot,y_0)) \to 0$ when $y_0 \to \infty$, uniformly for $P \in D'(\gamma) - E_\delta$, when $\gamma > 0$ and $\delta > 0$ are fixed. Thus

$$\lim_{y_0 \to \infty} I_3(\gamma,\delta,y_0) = 0 \; , \quad \gamma \text{ and } \delta \text{ fixed} \; . \tag{15.40}$$

To complete the proof of (15.29) let $\varepsilon > 0$ be given and use (15.37) to choose a $\gamma = \gamma_0 = \gamma_0(\varepsilon,g_0) > 0$ such that $I_1(\gamma,y_0) < \varepsilon/3$. Next use (15.39) with $\gamma = \gamma_0(\varepsilon,g_0)$ fixed to choose $\delta = \delta_0 = \delta_0(\varepsilon,g_0) > 0$ so small that $I_2(\gamma_0,\delta_0,y_0) < \varepsilon/3$. Both of these estimates hold uniformly for all $y_0 \ge 0$. Finally, choose $Y_0 = Y_0(\varepsilon,g_0)$ so large that $I_3(\gamma_0,\delta_0,y_0) < \varepsilon/3$ for all $y_0 \ge Y_0$. This is possible by (15.40). With these choices (15.32) implies that

$$\int_{R_0^2} |\rho(P,g(\cdot,y_0))|^2 \, dP < \varepsilon \text{ for all } y_0 \ge Y_0(\varepsilon,g_0) \; , \tag{15.41}$$

which proves (15.29) and therefore (15.24). Finally, to prove (15.25) one notes that if $f(\cdot,y_0) \in L_2^1(G)$ then $\omega(P)\ \hat{f}_-(P,y_0) \in L_2(R_0^2)$ and the preceding argument can be applied to this function. This completes the proof of Theorem 15.1.

<u>An Estimate of the Echo Waveform</u>. Under the hypotheses of Theorem 15.1 one has the estimate

$$\begin{aligned}\hat{h}_-^{sc}(\cdot,y_0) &= \hat{g}_-^{sc}(\cdot,y_0) - i\omega(\cdot)\ \hat{f}_-^{sc}(\cdot,y_0)\\ &= \hat{S}\ R\ \hat{g}(\cdot,y_0) - i\omega(\cdot)\ \hat{S}\ R\ \hat{f}(\cdot,y_0) + o(1) \qquad (15.42)\\ &= \hat{S}\ R\ \hat{h}(\cdot,y_0) + o(1)\end{aligned}$$

where $o(1) \to 0$ in $L_2(R_0^2)$ when $y_0 \to \infty$. Moreover, the mapping $\hat{h}_-^{sc} \in L_2(R_0^2)$ $\to e_0 \in L_2(R \times [0,\pi])$ defined by (15.12) is bounded with bound 1 [30, (2.84)]. It follows that

$$e_0(\tau,\theta) = \mathrm{Re}\left\{\frac{1}{(2\pi)^{1/2}} \int_0^\infty e^{i\tau\omega}(\hat{S}\,R\,\hat{h})(\omega\cos\theta,\omega\sin\theta,y_0)(-i\omega)^{1/2}\,d\omega\right\} + o(1) \qquad (15.43)$$

where $o(1) \to 0$ in $L_2(R \times [0,\pi])$ when $y_0 \to \infty$. Now

$$\begin{aligned}\hat{h}(p,q,y_0) &= \hat{g}(p,q,y_0) - i\omega(p,q)\ \hat{f}(p,q,y_0)\\ &= e^{iqy_0}[\hat{g}_0(p,q) - i\omega(p,q)\ \hat{f}_0(p,q)] \qquad (15.44)\\ &= e^{iqy_0}\ \hat{h}_0(p,q)\end{aligned}$$

and hence by (15.6)

$$\begin{aligned}(-i\omega)^{1/2}\ R\ \hat{h}(\omega\cos\theta,\omega\sin\theta,y_0) &= (-i\omega)^{1/2}\ \hat{h}(\omega\cos\theta,-\omega\sin\theta,y_0)\\ &= e^{-i\omega y_0 \sin\theta}(-i\omega)^{1/2}\ \hat{h}_0(\omega\cos\theta,-\omega\sin\theta) \qquad (15.45)\\ &= 2\ e^{-i\omega y_0\sin\theta}\ \hat{s}_0(\omega,-\theta)\ .\end{aligned}$$

Combining (15.43) and (15.45) gives

$$(15.46) \qquad e_0(\tau,\theta) = \text{Re}\left\{\left(\frac{2}{\pi}\right)^{1/2} \int_0^\infty e^{i\tau\omega}\, e^{-i\omega y_0 \sin\theta}\, \hat{S}(\omega)\, \hat{s}_0(\omega,-\theta)\, d\omega\right\} + o(1) .$$

Thus under the hypotheses of Theorem 15.1 the echo waveform is determined by the signal waveform, the S-matrix for the grating and the range parameter y_0, with an error that tends to zero in energy when $y_0 \to \infty$.

Pulsed Beam Signals. For many applications it is desirable to have a transmitter whose waveform $s_0(\tau,\theta)$ is sharply limited in both direction and frequency. The relation (15.6) shows that this could be achieved by choosing f_0 and g_0 such that supp $\hat{h}_0 = RK$ where $K \subset \mathcal{O}_{m,n}$ and m and n are suitably chosen. Of course, this condition cannot be satisfied with sources that are confined to a compact set, since if supp $\hat{h}_0$ is compact then $h_0(P)$ is analytic. However, it may be possible to choose f_0 and g_0 such that

$$(15.47) \qquad \hat{h}_0(p,q) = a(p,q) + b(p,q)$$

where

$$(15.48) \qquad \text{supp } a = RK \subset R\mathcal{O}_{m,n} ,$$

$$(15.49) \qquad \hat{s}_0^a(\omega,\theta) = \frac{1}{2}\,(-i\omega)^{1/2}\, a(\omega\cos\theta, \omega\sin\theta)$$

defines the desired waveform s_0^a and

$$(15.50) \qquad \|b\|_{L_2(R^2)} < \varepsilon .$$

If this transmitter design problem has been solved then the corresponding echoes will satisfy

$$(15.51) \qquad e_0(\tau,\theta) = e_0^a(\tau,\theta) + o_1 + o_2$$

where

$$(15.52) \qquad e_0^a(\tau,\theta) = \text{Re}\left\{\left(\frac{2}{\pi}\right)^{1/2} \int_0^\infty e^{i\tau\omega}\, e^{-i\omega y_0 \sin\theta}\, \hat{S}(\omega)\, \hat{s}_0^a(\omega,-\theta)\, d\omega\right\}$$

while

$$(15.53) \qquad \|o_1\|_{L_2(R\times[0,\pi])} < \varepsilon \text{ for all } y_0 \geq 0 \text{ , and}$$

$$\lim_{y_0\to\infty} \|\sigma_2\|_{L_2(R\times[0,\pi])} = 0 \ . \tag{15.54}$$

Angular Dispersion of Echoes from Gratings. The notation

$$\Gamma_\ell = \{P = (\omega\cos\theta, \omega\sin\theta) : \omega > 0 \text{ and } \alpha_\ell \le \theta \le \beta_\ell\} \tag{15.55}$$

will be used to denote the smallest sector such that $K_\ell = X_\ell(K) \subset \Gamma_\ell$, $-n \le \ell \le m$. The hypothesis $K = K_0 \subset \mathcal{O}_{m,n}$ implies that the sectors Γ_ℓ are disjoint and

$$\bigcup_{\ell=-n}^{m} \Gamma_\ell \subset R_0^2 \ . \tag{15.56}$$

Moreover, (15.48), (15.49) and (15.52) and Theorem 14.4 imply that one has

$$\text{supp } r^{-1/2} e_0^a(r-t,\theta) \subset \bigcup_{\ell=-n}^{m} \Gamma_\ell \tag{15.57}$$

for all $t > 0$. Thus, apart from the error terms in (15.51), the echo waveform is concentrated in the sectors Γ_ℓ. Note that in the case of a degenerate grating with Neumann (resp., Dirichlet) boundary condition one has $\hat{S} = 1$ (resp., $\hat{S} = -1$) and hence $r^{-1/2} e_0^a(r-t,\theta) = \pm r^{-1/2} s_0^a(r-t,\theta)$ has support in Γ_0. This is a well-known property of the specular reflection of a beam by a plane. In the case of a non-degenerate grating, where $\hat{S} \neq \pm 1$, one has only (15.57) and secondary reflected beams will appear in the sectors Γ_ℓ, $\ell \neq 0$. Their waveforms can be calculated explicitly using (15.52) and (14.87). They are distortions of the signal waveform $s_0(\tau,\theta)$ whose forms are determined by the scattering coefficients $c_\ell^-(\omega\cos\theta, \omega\sin\theta)$. This phenomenon of the angular dispersion of pulsed beams by diffraction gratings is the counterpart for transient wavefields of the phenomenon of the diffraction of monochromatic beams into the higher order grating directions.

References

[1] Agmon, S. Lectures on Elliptic Boundary Value Problems. Van Nostrand, 1965.

[2] Agmon, S. Spectral properties of Schrödinger operators and scattering theory. Ann. Scuola Norm. Sup. Pisa, Ser. IV, 2, 151-218 (1975).

[3] Alber, H. D. A quasi-periodic boundary value problem for the Laplacian and the continuation of its resolvent. Proc. Roy. Soc. Edinburgh, 82A, 251-272 (1979).

[4] Bloch, F. Über die Quantenmechanik der Electronen in Kristallgittern. Zeit. f. Phys., 52, 555-600 (1928).

[5] De Santo, J. A. Scattering from a periodic corrugated structure: thin comb with soft boundaries. J. Math. Phys., 12, 1913-1923 (1971).

[6] De Santo, J. A. Scattering from a periodic corrugated structure II: thin comb with hard boundaries. J. Math. Phys., 13, 336-341 (1972).

[7] De Santo, J. A. Scattering from a sinusoid: derivation of linear equations for the field amplitudes. J. Acoust. Soc. Amer., 57, 1195-1197 (1975).

[8] De Santo, J. A. Scattering from a perfectly reflecting periodic surface: an exact theory. Radio Sci., 16, 1315-1326 (1981).

[9] Dunford, N., and Schwartz, J. T. Linear Operators I. Interscience, 1957.

[10] Eckart, C. A general derivation of the formula for the diffraction by a perfect grating. Phys. Rev., 44, 12-14 (1933).

[11] Eidus, D. M. The principle of limiting absorption. Mat. Sb. 57, 13-44 (1962) = AMS Transl. (2) 47, 157-191 (1965).

[12] Ikebe, T. Eigenfunction expansions associated with the Schrödinger operator and their application to scattering theory. Arch. Rational Mech. Anal. 5, 1-34 (1960).

[13] Kato, T. Perturbation Theory for Linear Operators. Springer-Verlag, 1966.

[14] Lions, J. L., and Magenes, E. Non-homogeneous Boundary Value Problems and Applications I. Springer-Verlag, 1972.

[15] Lyford, W. C. Spectral analysis of the Laplacian in domains with cylinders. Math. Ann., 218, 229-251 (1975).

[16] Narasimhan, R. Several Complex Variables. CUP, 1971.

[17] Petit, R., Ed. Electromagnetic Theory of Gratings, Topics in Current Physics. Vol. 22. Springer-Verlag, 1980.

[18] Rayleigh, Lord (J. W. Strutt). The Theory of Sound. 2nd Ed. Vol. 2, 89-96, Dover, 1945.

[19] Rayleigh, Lord (J. W. Strutt). On the dynamical theory of gratings. Proc. Roy. Soc., Ser. A., 79, 399-416 (1907).

[20] Rellich, F. Über das asymptotische Verhalten der Lösungen von $\Delta u + ku = 0$ in unendlichen Gebieten. Jber. Deutschen Math. Verein, 53, 57-64 (1943).

[21] Schulenberger, J. R. and Wilcox, C. H. Eigenfunction expansions and scattering theory for wave propagation problems of classical physics. Arch. Rational Mech. Anal., 46, 280-320 (1972).

[22] Shenk, N. A. Eigenfunction expansions and scattering theory for the wave equation in exterior regions. Arch. Rational Mech. Anal., 21, 120-150 (1966).

[23] Shizuta, Y. Eigenfunction expansions associated with the operator $-\Delta$ in the exterior domain. Proc. Japan Acad., 39, 656-660 (1963).

[24] Steinberg, S. Meromorphic families of compact operators. Arch. Rational Mech. Anal., 31, 372-379 (1968-69).

[25] Stroke, G. W. Diffraction Gratings. Handbuch der Physik XXIX. Springer-Verlag, 1967.

[26] Uretsky, J. L. The scattering of plane waves from periodic surfaces. Ann Phys., 33, 400-427 (1965).

[27] Weinstein, L. A. The Theory of Diffraction and the Factorization Method. Golem, 1969.

[28] Wilcox, C. H. Initial-boundary value problems for linear hyperbolic partial differential equations of the second order. Arch. Rational Mech. Anal., 10, 361-400 (1962).

[29] Wilcox, C. H. Scattering states and wave operators in the abstract theory of scattering. J. Functional Anal., 12, 257-274 (1973).

[30] Wilcox, C. H. Scattering Theory for the d'Alembert Equation in Exterior Domains. Lecture Notes in Mathematics. Vol. 442. Springer-Verlag, 1975.

[31] Wilcox, C. H. Spectral and asymptotic analysis of acoustic wave propagation. Boundary Value Problems for Linear Evolution Partial Differential Equations. Reidel, 1977.

[32] Wilcox, C. H. Theory of Bloch Waves. J. d'Anal. Math., 33, 146-167 (1978).

[33] Wilcox, C. H. Sonar echo analysis. Math. Meth. in the Appl. Sci., 1, 70-88 (1979).

[34] Wilcox, C. H. and Guillot, J. C. Scattering theory for acoustic diffraction gratings - preliminary report. Notices AMS, Vol 25, A356 (Jan. 1978).

Index

Applied Mathematical Sciences

36. Bengtsson/Ghil/Källén: **Dynamic Meterology: Data Assimilation Methods.**
37. Saperstone: **Semidynamical Systems in Infinite Dimensional Spaces.**
38. Lichtenberg/Lieberman: **Regular and Stochastic Motion.** (cloth)
39. Piccinini/Stampacchia/Vidossich: **Ordinary Differential Equations in R^n.**
40. Naylor/Sell: **Linear Operator Theory in Engineering and Science.** (cloth)
41. Sparrow: **The Lorenz Equations: Bifurcations, Chaos, and Strange Attractors.**
42. Guckenheimer/Holmes: **Nonlinear Oscillations, Dynamical Systems and Bifurcations of Vector Fields.**
43. Ockendon/Tayler: **Inviscid Fluid Flows.**
44. Pazy: **Semigroups of Linear Operators and Applications to Partial Differential Equations.**
45. Glashoff/Gustafson: **Linear Optimization and Approximation: An Introduction to the Theoretical Analysis and Numerical Treatment of Semi-Infinite Programs.**
46. Wilcox: **Scattering Theory for Diffraction Gratings.**
47. Hale et al.: **An Introduction to Infinite Dimensional Dynamical Systems – Geometric Theory.**
48. Murray: **Asymptotic Analysis.**